Hybrid Museum Experiences

Hybrid Museum Experiences

Theory and Design

Edited by
Annika Waern and
Anders Sundnes Løvlie

Routledge
Taylor & Francis Group
LONDON AND NEW YORK

The publication of this book is made possible by a grant from the European Union's Horizon 2020 research and innovation programme under grant agreement No 727040 (the GIFT project).

First published in 2022 by Amsterdam University Press Ltd.

Published 2025 by Routledge
4 Park Square, Milton Park, Abingdon, Oxon OX14 4RN
605 Third Avenue, New York, NY 10158

Routledge is an imprint of the Taylor & Francis Group, an informa business

ISBN: 9789463726443 (hbk)
ISBN: 9781041181156 (pbk)
ISBN: 9781003697657 (ebk)
NUR 670

Cover illustration: Charlie Johnson

Cover design: Coördesign, Leiden

DOI: 10.5117/9789463726443

Table of Contents

Craft

Coda

Preface

Kevin Bacon and Nikita Mathias

The digitization of museum collections has led to millions of digital representations of objects to be experienced, shared and discussed by online visitors. In most cases, these digital copies live a life of their own, disconnected from the real objects they represent. 'Hybrid museum experiences' seek to bring together the original object and digital technology to facilitate visitor experiences that are meaningful, personal and inclusive. Combining the emotional and intellectual force of objects in museum spaces with the participatory, narrative and immersive potential of digital, hybrid experiences aim at reconfiguring the relationship between the visitor, the exhibit and the museum that surrounds them.

This book is an ideal reading for anyone interested in museums who is open to an ambitious yet deeper and more nuanced approach to digital technology. Although it is mostly written by academic researchers, it has been informed by work with a variety of museum practitioners, including the authors of this preface.

We met as part of the action research strand of the GIFT project, which brought together museum professionals from five different countries. Although we represented a variety of museums differing greatly in scope and size, we recognized several shared challenges. How do we get more of our colleagues working with digital technologies? How do we create genuinely audience-focused experiences? How do we embed experimental practice into our organizations?

These are thorny questions which resist simple solutions and require a good deal of honesty and self-awareness to try and answer. The workshops supported by GIFT created a safe space for reflection, discussion and using design thinking tools to work collectively toward common goals. Moreover, by establishing a dialogue between different museum sectors, country borders and various disciplines (design, academia, cultural heritage, visual arts, computer science) it equipped those discussions with a rich conceptual vocabulary that helped all the participants to rethink what a museum can mean to its visitors.

Waern, A. and A. Sundnes Løvlie (eds.), *Hybrid Museum Experiences: Theory and Design*. Taylor & Francis Group, 2022
DOI 10.5117/9789463726443_PRE

In addition to the action research workshops described in Chapter 6 both MUNCH and Royal Pavilion & Museums supported the prototyping of some of the larger projects featured in this book. The *Gift* app (see Chapter 3) developed by GIFT partner Blast Theory was iteratively tested in Brighton Museum over three years before becoming part of the visitor experience there, and at MUNCH in 2019. MUNCH also collaborated with GIFT partner NextGame on its *Sensitive Pictures* prototype and Brighton Museum contributed to an app developed within GIFT, the *One Minute* app, from content creation to in-gallery deployment.

Some of these projects are described in more detail in the book, and even more are represented at the accompanying website. It is worth noting that our museums were far from passive in the development and testing processes and had an active role in shaping the experiences. In the case of the *Gift* app we used this project to try to solve subtly different problems for our respective museums. For Royal Pavilion & Museums it was an opportunity to experiment with a way of bringing coherence to the eclectic collections displayed in Brighton Museum, by enabling our visitors to create and share personal interpretations. Our role in working with Blast Theory was primarily to help define the affordances of the museum by making the artists aware of where barriers to uptake may exist. Often those barriers were physical, such as the counter-intuitive layout of the galleries, but they were also human: How do you motivate front of house staff to understand and promote a new digital experience? The openly collaborative approach practised by Blast Theory stood in marked contrast to the solutionism sometimes practised by commercial technologists, where a product is designed around a narrow problem ('how to enable visitors to learn more about the artworks?') without adequately considering local factors such as the admission process and the existing behaviour patterns of visitors.

At MUNCH, Blast Theory's *Gift* app had to function in an entirely different environment. In contrast to the diverse collection of Royal Pavilions & Museums, we explored the possibilities of the app within the context of a temporary exhibition dedicated to the art of Edvard Munch and a few other artists represented in our collection. But the exhibition was also about gifting. Since our collection is overwhelmingly based on donations by Edvard Munch and others, we wanted to explore what it means to receive, give and care for a gift. The *Gift* app gave our visitors a tangible and personal gifting experience and, hopefully, made them think about gifting as a social practice. However, we also learnt that gifting is a complicated and fragile matter. What is better? To receive or to give a gift? Our test runs showed that using the *Gift* app to create a gift for someone is more rewarding than receiving a gift from the museum. Yet, for the visitor, who does not know how rewarding the experience can be, the prospect of making and sending a gift may not be reason enough to give the app a try. This was a valuable lesson when experimenting with different types of materials and call-to-actions for introducing visitors to the experience. The visitors' motivations to use or not use the *Gift* app were more complex than we anticipated, especially when catering to different target groups such as primary school pupils, tourists, or young adults from Oslo.

Both of our museums have learned a great deal from the GIFT project, above and beyond the component parts we were involved in. We have seen the productivity of iterative processes and scalable prototyping; how museums can collaborate with small and agile creative companies through facilitating detailed user testing; and the power of good and accessible storytelling allied with short feedback loops. Even some of the relative failings of these projects have proven valuable: Wrestling with the challenges of introducing visitors to these digital experiences has helped shape the 'bring your own device' tactics now used by Royal Pavilion & Museums. As for the application of Sensitive Pictures at MUNCH, several shortcomings in the attempt to address the visitor in a personal yet non-intrusive way eventually led to hybrid experiences that established meaningful and emotional connections with Edvard Munch's art. This was achieved through a series of prototypes and a *minimum viable product* approach, starting off with a theatre workshop and plain prototypes, which were experienced as being too intrusive, too intense and too serious by many users. Based on these relative failures, we were able to design an experience that struck the right balance between intimacy and playfulness.

We certainly see the opportunities for hybrid experiences in museums: The potential for inclusion and interactivity; to allow for freedom of art experience in otherwise restricted physical museum spaces; to open up heritage spaces for personal and storytelling-based experiences. Yet our participation also made us realise the importance of user-centeredness, iteration and testing, service design thinking, not only for hybrid experiences but for all the kinds of experiences, products and services a museum provides.

It is also worth noting that this book is published at a time when museums need to become much smarter in how they use digital. The increased expectations of funders for museums to use more digital technology has coincided with a period of austerity in many countries following the financial crash of 2008. This has often resulted in funding becoming focused on one-off flagship projects, short-term R&D initiatives with no support to scale up to full deployment, and an over-reliance on proprietary social media platforms that offer ephemeral engagement at best. The findings of this book strongly indicate that such short-term thinking and its correlating funding schemes are bound to fail if not backed up by nurturing the establishment of agile and human-centred production processes as an essential in-house component of museums.

Witnessing the rise of the *experience society* and *economy*, museums are further pressured to compete with a multitude of experiential offers such as escape rooms, multiplex cinemas, streaming platforms, VR arcades, immersive theatre performances and digital playgrounds, to name but a few. What these offers strikingly demonstrate is that only those who know and cater to the needs and challenges of their users succeed. Museums can simply not afford to ignore their visitors anymore, especially when it comes to costly digital products. This aspect of human-centeredness must eventually transform every element of the museum visit, thereby also slowly eroding well-established notions and conventions that dictate what to expect and how to behave inside a museum. The app-based hybrid experience *Never let me go*, which resulted from the GIFT project (see Chapter 4), empowers the visitor to playfully explore and challenge the boundaries of museum conventions. Digital assets as such can become powerful tools to make museums more inclusive and relevant for broader parts of society.

At the time of writing (November 2020), many museums have been forced to close their doors during the Covid-19 pandemic and the subsequent economic shock is likely to place further financial strain on the sector over the next decade.

Museums need to become much more agile and efficient in how they use digital technology, demonstrating genuine impact as a return on investment. By introducing a set of open source tools and, more importantly, introducing fresh and radical ways of rethinking the museum experience, this book is both timely and essential for the sector.

About the Authors

Kevin Bacon is Digital Manager at the Royal Pavilion & Museums Trust, a charity managing five museums in the city of Brighton & Hove. With previous experience working in both front of house roles and as a curator of its photographic collections, he became the service's first digital lead in 2011. He holds Master's degrees in political philosophy and digital media and is a Visiting Researcher at the University of Brighton.

Nikita Mathias is a senior concept developer at MUNCH, where he creates digital and analogue visitor experiences, does research and works on publications. His background lies in art history, media studies and aesthetics, and he holds a PhD from the University of Tübingen, Germany, on the topic of the visual history of natural disasters. In addition, he spent years working as a journalist and at various cultural institutions. Born and raised in Germany, he has been living in Oslo for several years.

Concepts

1. Introduction

Anders Sundnes Løvlie and Annika Waern

Abstract

Introduces the concept of hybrid museum experiences, why it is timely, and presents the major theme of the book: How to make museum experiences more personally and socially meaningful through the use of technology.

Keywords: Hybrid Museum Experiences; Personalisation; Interpersonalisation; Museum Technology

You are walking the corridors of the National Museum in Belgrade. Strolling through majestic exhibition halls you view the collections of historic artefacts and artworks. On the side of a painting depicting a ferocious medieval battle scene ('Study for the painting Furor Teutonicus', by Paja Jovanović) you notice a label written with a peculiar font, saying: '#war'. You scan the label with your phone, and suddenly the phone's camera screen projects a virtual object in front of you, coming from a much more recent era: A red triangular warning sign that reads 'MINES'. Below it, a text appears:

> *Serbian Sarajevo, winter of 1995, Dayton agreement just signed. I'm twenty-two years old and filming a documentary film. The surrounding streets all barricaded, everything is ruined, abandoned, the buildings are riddled with shrapnel. The street before me is empty, without a living creature in sight. I spot a single rope across it with a red sign hanging that reads 'mines'. I walk up to it and take it down, without fear and without logic I decide to keep it as a souvenir. I keep the MINES sign as an anti-war protest sign, in preparation, because there will be more of them to come.*

This is a part of *Your Stories*, the result of a collaborative project between the National Museum in Belgrade and the Serbian design agency NextGame, in which citizens were invited to virtually 'donate' mundane objects of great

Waern, A. and A. Sundnes Løvlie (eds.), *Hybrid Museum Experiences: Theory and Design*. Taylor & Francis Group, 2022
DOI 10.5117/9789463726443_CH01

personal importance to them. The objects were scanned using photogrammetry, and were exhibited as virtual 3D models along with a short text explaining the significance of the object to the person who donated it. The collection of virtual objects and the way they were associated with physical museum objects set up a parallel digital exhibition – a 'people's museum' accessible within the physical space of the museum and aligned with the permanent exhibition.

The *Your Stories* project is an example of what we in this book call hybrid experiences: Designs that use new technologies to augment, expand or alter the physical experience of visiting the museum. We will discuss several other examples throughout the book. One will be a phone app that allows visitors to record their personal reflections on a museum collection and send them as a gift to somebody, who in their turn can experience the gift on site in the museum, as a personal guide. Another example that will be discussed in detail is a game in which one museum visitor controls the movements of another visitor as an 'avatar', to create a playfully personalized experience of the museum.

What do these examples have in common? First, all use technology – often smartphone-based – with the aim of adding a digital dimension to the museum visit. The use of digital technology in the museum sector has often been viewed as an alternative to the physical museum – sometimes manifested in the idea of the 'virtual museum', and at the time of writing reflected in the proliferation of online interfaces to cultural heritage, museum APIs[1] and the Open GLAM[2] movement[3]. Hybrid museum experiences break from this trend, in that they are closely integrated within a physical museum visit. Even those that are smartphone and internet based (and thus could in principle be accessed anywhere, anytime) are designed to be experienced in the context of a physical museum exhibition. As such they place themselves in a middle position (a hybrid) between the 'traditional physical museum experience', and a purely virtual experience. Furthermore, unlike many other uses of technology inside museums, hybrid museum experiences are not digital experiences designed to be experienced in isolation – such as a stationary information screen or kiosk – but rather in close relation to the physical space and exhibit. In the example above, this ambition can be seen in the way the digital content has been carefully chosen to relate to the physical exhibits it is attached to.

1 Application Programming Interface. See for instance https://pro.europeana.eu/page/apis.
2 https://openglam.org
3 Schweibenz, 'The Virtual Museum: An Overview of its Origins, Concepts, and Terminology'.

The design aim for hybrid museum experiences, of integrating digital technology closely with the museum space and the museum visit, means that hybrid museum visits need to strike other difficult balances – they need to be hybrid in more than one respect. In particular, hybrid museum experiences need to take into account how museum visits are performed. Museum technology falls on a scale between presenting content to single users – such as the ubiquitous audio/multimedia guide applications – towards broadcasting (or 'sharing') museum content to broad audiences through social media, serving the museum's purposes of marketing. However, both of these extremes are problematic: On the one hand, museum visits are usually social activities where visitors are interested in doing something together, which means that experiences that can only be used by single visitors will fail to accommodate the visitors' needs and interests. On the other hand, sharing content through social media may reach a broad audience, but also risks becoming incorporated into social media logics leading to mostly shallow interactions such as 'likes' and brief comments. There is great potential for creating hybrid museum experiences that explore the space in between these extremes, taking into account how museum visits are *interpersonal*, shared with close friends or in groups where the museum visit serves a function of strengthening social ties. Hybrid museum experiences should be designed to be shared – not with a large audience on social media, but rather with one or a few people that are 'special': Family, friends and loved ones. This is another way in which these experiences are hybrid: They are social, while still highly personal. In *Your Stories* the interpersonal dimension is explored through the personal objects and stories which have been donated by people who themselves are visitors of the museum, and which now can be shared with other visitors. As we will further elaborate in Chapter 5, the curator's aim was also to create a more intimate connection with the historical artefacts on display, through their association with the more current, and sometimes mundane, stories told by donors.

Finally, hybrid experiences need to strike a balance in integrating both the museum and the visitor perspectives. Hybrid museum experiences are designed to foreground visitor perspectives, they open up for active exploration of museums and they will very often integrate contributions from visitors (such as in the *Your Stories* example). However, they need to do so without losing track of the importance of curatorship and the way museums create meaning. Thus, they become hybrid in the sense that they explore the design space between two concepts often applied to digital media: Participation and curation. In *Your Stories*, we see this balance in how the call for objects and stories was open to any and all visitors, while the selection of what objects to include in the exhibition and which exhibits to associate them with, was a deliberative process in the hands of the curators.

To sum up, the experiences we discuss in this book are hybrid in relation to multiple dimensions of the museum experience:

- The physical and the digital
- The personal and the social
- The museum and the visitor

This means that while technology is used to make these experiences possible, technology is not at the centre. Our examples mostly use market-ready technologies. Instead, it is the human *experience* and the meaning-making processes that are ongoing during the museum visit that are at the centre.

We – the authors of this book – are researchers and designers working in the fields of experience design, human-computer interaction, play and media studies. We believe that hybrid museum experiences may be used to offer engaging experiences to visitors, to allow the museum to reach out to new audiences, and even challenge the boundary between audience and participants in a collective reimagining of what museums can and should be.

Why hybrid museum experiences?

Museums are changing. Once, museums were seen primarily as repositories housing valuable and venerable artefacts; today they are increasingly seen as places of experience and dialogue, they are multi-voiced, story-oriented, open for provocation and reflection. Over recent years, museum institutions have increasingly shifted their focus from highlighting physical collections to highlighting stories and experiences they can share with their audiences.[4]

In this book, we focus on how technology can better involve the museum audience, so that visitors and their preconceptions and experiences become part of defining what museum artefacts can mean to them and their group. Museum visitors *will* create their own meaning of an exhibit (and sometimes share it), even though the museum and curators control what is displayed and how. A main challenge for museums is: How can they support visitors in their meaning-making process, letting them be active participants in their visit? Hybrid museum experiences can be both challenging and open

4 Hooper-Greenhill, 'Changing Values in the Art Museum'; Hooper-Greenhill, *Museums and the Interpretation of Visual Culture*; Simon, *The Participatory Museum*.

for appropriation, and thereby allow for a rich variety of personal meaning-making processes.

The technologies needed to create hybrid experiences have existed for some time, and have dominated much work with digital museum experiences in recent years. At the time of writing this book, in particular Augmented Reality (AR) and Virtual Reality (VR) have been in vogue for some years, driven by commercial products like the AR-based games *Pokémon GO* and *The Walking Dead: Our World*, as well as the increasing availability of consumer-targeted VR headsets like Oculus Rift and HTC Vive. Of particular relevance for this book is the concept of Mixed Reality, which has been described as a subset of Virtual Reality and contains a continuum of ways to combine real and virtual input.[5] Mixed Reality has been used for artistic and entertainment purposes for a long time, and many of the authors of this book have been central in that development through numerous research projects throughout the last decades.[6]

However, the term 'hybrid' is in some sense deliberately vague. It indicates a mix of other elements – in this case, some elements of digital technology combined with some elements of 'analogue' physical exhibits – but doesn't clearly specify which technologies are used, or in what kind of physical setup. To some extent, the perspective we are offering is technology-agnostic: We give primacy to designing *experiences*, and employ the technologies that are available to deliver these experiences. However, through the chapters of this book we will present some particular tools and solutions that we find particularly promising to explore in a museum context.

Experience design

What does it mean to focus on experiences rather than technology – and why is that a good idea? After all, digital museum experiences do rely on technology. However, it might be useful to contemplate more precisely what we mean when we talk about technology. Are we talking about hardware devices such as smartphones, interactive displays, virtual reality headsets? Or are we more interested in the software – such as apps, games, or chatbots? Or is there some other dimension we should be looking at?

The media scholar Gunnar Liestøl has suggested that the development of digital media takes place not just on the levels of hardware and software,

5 Milgram and Kishino, 'A Taxonomy of Mixed Reality Visual Displays'.

6 Benford and Giannachi, *Performing Mixed Reality*; Montola, Stenros, and Waern, *Pervasive Games: Theory and Design*.

but also on a third level he calls 'meaningware'.[7] This concept refers to genres of communication, as they are reimagined and reshaped using the new technologies. Liestøl's central observation is that the development seems to go at different speeds in the different levels: While hardware develops at an astonishing rate and new software often also seems to develop bewilderingly fast, innovations in digital genres – the new aesthetics and modes of meaningful engagement enabled by digital technology – tend to be slower and further apart. The state of digital technology in museums seems to support this view. At the level of hardware and software, museums have for a long time embraced digital technology and have spent considerable resources digitizing collections and building digital infrastructures. However, innovation in the design of visitor experiences that use digital technology lags behind.

We can see this for example in how the museum sector has adapted to smartphones. Museums have started to take into account the most straightforward technical capabilities of smartphones, e.g. through offering museum guides as downloadable apps. But it has been harder for museums to adapt to the way in which visitors also arrive with their smartphone habits, conventions and expectations. While some museums have begun to invite visitors to take photos and share museum content through social media (e.g., through Instagram competitions), the very act of photographing within the museum remains problematic both for preservation and legal reasons. At the time of writing, museums have started to instead use app technology to develop personalizable and versatile museum guides, and mobile apps such as Smartify, Vizgu or Magnus use image recognition to this purpose.

Digital technology can offer so much more. Museums have so far only begun to explore the vast potential for meaningful experiences enabled by digital technologies including the smartphone. Typical examples of hybrid technologies that now are entering the museum sector include augmented and virtual reality, tangible interactions, and various targeted devices – even chatbots such as *Send Me SFMOMA*.[8] These formats offer the potential for experiences that are very different from ordinary museum experiences, including pervasive games, alternative narratives, and deeply personal immersive experiences.

Digital technology also brings additional complexity to the museum context, and adds new responsibilities. Museums face a bewildering range

7 Liestøl, 'PowerPoint: Beyond Hardware and Software'.

8 Chan and Cope, 'Strategies against Architecture'; Mollica, 'Send Me SFMOMA'.

of technologies, each with its own benefits but also posing new demands on museum personnel when it comes to making creative use of these, commissioning new forms of experiences, and maintaining the results. The pitfalls are many: Technologies may fail, they may require too much in resources or upkeep, the completed designs may not do quite what was initially promised, designs may fail to meet the complex and sometimes conflicting demands of curators, marketers, educators and IT departments – or they may simply fail to engage visitors.

Developing technologically mediated visitor experiences is a 'wicked problem'[9] of the kind that designers specialize in solving, by gathering contributions from multiple stakeholders and prioritising human experience over technological paradigms. However, far too often this problem is approached by putting technology first, and engineering systems and apps that often do not end up meeting the actual needs of museums or visitors. In this book, we explore the opportunities and challenges brought about by new museology and technology through the lens of hybridity. This lens allows us to acknowledge that solutions do not lie solely with the technology, but in the careful crafting of the meeting between technology, the physical museum, museum professionals and visitors.

GIFT

The theory, cases, methods and tools presented in this book were all developed within the EU-funded research project GIFT: A large and cross-disciplinary research project involving artists, designers, curators, museum educators, computer scientists and a high number of museums in Europe and the US. All of the authors of this book were participants in the project. The book also draws on a wide range of research and examples based on the authors' decades of collective experience as designers and researchers in this area.

When the GIFT project was originally devised, in the winter of 2015–2016, much of the museum world was in the throes of an ongoing VR hype. At the time, it seemed like every other research project involving museums and technology were dedicated to re-creating a museum experience inside the virtual world of a VR headset. But as already discussed, visits are seldom isolated: Most people come with someone else, whether this is their significant other, a whole family, a group of friends, or a class of school children.

9 Buchanan, 'Wicked problems in design thinking'.

The VR trend created problems from this perspective: When you put on a VR headset you are cut off from the people around you and immersed in a purely virtual world (while outside, your friends are impatiently waiting for their turn, making fun of the strange gestures you are doing, oblivious to your surroundings). We decided to instead shift our perspective towards human practices and experiences that in everyday life rely on and reinforce the social ties within groups. The social practices that we selected to explore were *gifting* and *play*, perspectives that will be detailed further in Chapters 2, 3 and 4.

In the GIFT project, we have collaborated with organizations such as Culture24 and Europeana, that have worked extensively over a long time to understand what it takes to build the digital capacity of a museum. Furthermore, we have worked with a large number of museums in Europe and the US to explore challenges and possible improvements to the way they work with digital and hybrid experiences. These include:

- San Francisco Museum of Modern Art, USA
- TATE Modern, UK
- The Munch Museum, Norway
- ARKEN Museum of Modern Art, Denmark
- The National Gallery of Denmark
- Royal Albert Memorial Museum & Art Gallery, UK
- Brighton Museum & Art Gallery, UK
- CAOS Centro Arti Opificio Siri, Italy
- Center for Studies of Holocaust and Religious Minorities, Norway
- The National Museum in Belgrade, Serbia
- The National Videogame Arcade, UK
- Danish Museum of Science & Technology, Denmark
- Derby Silk Mill, UK
- Tyne & Wear Archives & Museums, UK
- The Museum of Yugoslavia, Serbia
- University of Nottingham Museum of Archaeology, UK
- The National Museum of Photography, Denmark
- The Danish Architecture Center, Denmark

Working with these museums has provided us with important insights into the everyday difficulties faced by museum professionals, in trying to integrate technologically mediated experiences with their collections. It has also provided a testing ground and reality-check for methods, tools and concrete designs coming out of the project. Throughout the project and through the collective sharing of experiences, the group of museums developed ways of working that can be replicated in other museums. In later chapters in this book we share some tools and recommendations, that can be used by museum professionals and others working with museum organizations in order to improve their efforts with hybrid experiences.

The structure of this book

This book is divided into four main sections called concepts, cases, craft, and coda. The concepts section aims to give an overview of the main themes of the book and build a theoretical framing for both the concept of hybrid museum experiences, and the design approach advocated throughout the book. In the following chapter, the concept of Hybrid Museum Experiences is discussed in further depth, and related to previous research in this area.

The second section presents three case studies of designs that were created as part of the GIFT project, and tried out in six different museums in the UK, Denmark, Norway and Serbia. Chapter 3, called 'The *Gift* App – Gifting Museum Experiences' presents the design and deployment of a web app that invites museum visitors to turn their visit into a hybrid experience that can be digitally 'wrapped' as a gift and sent to someone they love – as if they were making a mixtape, only with objects from a museum. The chapter explores how gifting practices can be brought into museum visits through a hybrid museum experience, to make them more personally and interpersonally meaningful.

Chapter 4, called '*Never let me go*: Social and Introspective Play', presents a case study that directly addresses the way in which museum visits are rarely solitary experiences as visitors come in pairs or in groups, and their main reason to visit is typically to spend time with the people they come with. Museums tend to put great emphasis on the objects they have on display, and while they do want visitors to have an enjoyable (and possibly also social) experience, they also want to direct the visitors' attention towards the exhibition. The chapter explores this dilemma through the design of a playful museum experience called *Never let me go*, which aims to facilitate an experience that is at one and the same time social, while also directing the participants towards an introspective encounter with the museum collection.

The final case study Chapter 5, called '*Your Stories*: The Lifecycle of a Museum Experience', is a more in-depth account of the *Your Stories* case presented in the vignette at the start of this chapter. It looks into the potential for participatory curatorship and how the roles of the audience, the museum and the technology developer were balanced in creating the experience. It also deals with another challenge for hybrid museum experiences: How can the museum ensure that a hybrid experience works well for the people it is intended for – and how can they make sure that it continues to work well, throughout the time that the system is in place? Through a study of this augmented reality installation in the National Museum in Belgrade, the chapter discusses the many practical and organizational challenges involved in developing and maintaining hybrid museum experiences.

In the craft section, we present a number of tools and methods for working with hybrid museum experiences. The first two of these chapters deal with the initiation of the design process. In Chapter 6, we discuss how museums can use action research as a method for 'reflective practice', which is a central concept in design research. We argue that developing a culture of internal reflective practice is vital for dealing with and getting the most out of the digital opportunities and challenges that museums are facing. Action research may be one productive way of fostering such a culture. The methods presented in the two following chapters are particularly well suited for use within the context of action research processes, even if they also can be used on their own.

Chapter 7 is called 'Sensitizing to Theory' and addresses a well-known problem in design research: How can relevant insights from academic theory help to inform practical design processes? The chapter presents two methods for doing this: Through introducing a theoretician as a dedicated role in a designer team, and through role-playing scenarios that can help designers and museum professionals to get a shared, embodied understanding of theories and concepts that are crucial for their joint design process, and that can help further their discussions during the same.

Chapter 8, 'Ideation Tools for Experience Design', presents insights from the practical experiments conducted as part of the action research project in GIFT. As part of this process a number of tools were developed and tested, to support the process of developing ideas, from the very early brainstorming phase and onwards to a critical examination of ideas, and early involvement of different stakeholders in the organization.

Chapter 9, 'Data Driven Visitor Experiences', discusses how data collected through digital tools can be used to develop a deeper understanding of visitor behaviour and needs, both during design and when a hybrid experience is in place in the museum. It is however not easy to make sense of data, even when one has access to it: This chapter delves into a variety of forms of data collection and visualization that present useful insights into visitor behaviour and experiences.

Chapter 10, 'Evaluating Hybrid Experiences', is a chapter about how hybrid museum experiences can be studied and evaluated. Hybrid museum experiences are put in place for a range of reasons, and evaluations help illuminating if, and how, the goals of a particular design were met. While evaluations sometimes only tell 'what was already known', they help put hard facts and figures to such imprecise impressions. And sometimes, they surprise us, uncovering effects and issues that were completely unknown.

In the final section called 'coda', the concluding chapter of the book returns to the critical aspects of hybridity, to discuss the opportunities and risks related to the design and employment of hybrid museum experiences. We highlight how the introduction of a hybrid museum experience, especially when it is one that is designed to shape the entire museum visit, challenges and sometimes redefines the essence of museums: What a museum *is*, and what the museum visit is about. The chapter discusses how the introduction of technology may remediate, reframe, or sometimes entirely restage the museum experience and how this can be seen as at the same time a challenge to the museum's identity, and an opportunity for museums to reinvent themselves.

The book forms a part of the resources that have come out of the GIFT project. More resources have been made available through the website https://gifting.digital. On the website, you can find the tools we are presenting in the book and more, including all the documentation that is needed to use the tools, the source code of software tools, case studies from practical experiments, scholarly publications, and other materials that might be useful for museum professionals and designers who wish to put these tools to use in practice. All of the tools offered on the website are open source and can be used freely and adapted to your own needs (even if you wish to use them for commercial purposes). Efforts have been made to make tools useful even for museums with few resources and limited technical capacity.

Bibliography

Steve Benford, and Gabriella Giannachi, *Performing Mixed Reality* (Cambridge, MA, London, U.K.: The MIT Press, 2011)

Richard Buchanan, 'Wicked problems in design thinking', *Design issues*, 8.2 (1992), 5–21

Sebastian Chan, and Aaron Cope, 'Strategies against Architecture: Interactive Media and Transformative Technology at the Cooper Hewitt, Smithsonian

Design Museum', *Curator: The Museum Journal*, 58.3 (2015), 352–368 <https://doi.org/10.1111/cura.12118>

Eilean Hooper-Greenhill, 'Changing Values in the Art Museum: Rethinking Communication and Learning', *International Journal of Heritage Studies*, 6.1 (2000), 9–31 <https://doi.org/10.1080/135272500363715>

Eilean Hooper-Greenhill, *Museums and the Interpretation of Visual Culture*, Museum Meanings, 4 (Oxon, New York: Routledge, 2000)

Gunnar Liestøl, 'PowerPoint: Beyond Hardware and Software', in *Inside Multimodal Composition*, ed. by Andrew Morrison (New York: Hampton Press, 2009), pp. 71–93

Paul Milgram, and Fumio Kishino, 'A Taxonomy of Mixed Reality Visual Displays', *IEICE TRANSACTIONS on Information and Systems*, E77-D.12 (1994), 1321–1329 <https://search.ieice.org/bin/summary.php?id=e77-d_12_1321> [accessed 25 February 2019]

Jay Mollica, 'Send Me SFMOMA', *San Francisco Museum of Modern Art*, 2017 <https://www.sfmoma.org/read/send-me-sfmoma/> [accessed 23 September 2019]

Markus Montola, Jaakko Stenros, and Annika Waern, *Pervasive Games: Theory and Design* (Boca Raton, FL: CRC Press, 2009)

Donald A. Schön, *The Reflective Practitioner: How Professionals Think in Action* (Aldershot, UK: Ashgate, 1983)

Werner Schweibenz, 'The Virtual Museum: An Overview of its Origins, Concepts, and Terminology', *The Museum Review*, 4.1 (2019), 2–28

Nina Simon, *The Participatory Museum* (Santa Cruz, California: Museum 2.0, 2010) <http://www.participatorymuseum.org/> [accessed 27 January 2016]

About the Authors

Anders Sundnes Løvlie does research on the intersection of design research and media studies, focusing in particular on experience design, locative media and play. Anders was the coordinator for the GIFT project, and has been involved in a number of design projects involving museums like the Munch Museum, Danish Architecture Center, the National Gallery of Denmark, Brighton Museum and the Frederiksberg Museums. Author of the book *The Rhetoric of Persuasive Games: Freedom and Discipline in America's Army*. Leads the research group on Media, Art and Design (MAD).

Annika Waern is a 'research through design' academic who has been researching technology-supported physical play and games for about fifteen years. Currently, she is conducting research in the areas of hybrid play in museums, children's play in outdoor settings, and circus training to foster proprioceptic skills. Waern has authored over 100 academic articles and book chapters, and been editor of 2 books and several journal special issues. She is a DiGRA (Digital Games Research Association) distinguished scholar and a HEVGA (Higher Education Video Game Alliance) fellow.

2. Hybrid Museum Experiences

Anders Sundnes Løvlie, Annika Waern, Lina Eklund, Jocelyn Spence, Paulina Rajkowska and Steve Benford

Abstract
The concept of 'Hybrid museum experiences' is inspired by New Museology, and looks at how museum experiences can be made more meaningful by giving visitors agency in their museum experience. The chapter emphasises how museum experiences are inherently social, and looks at how hybrid museum experiences can be made meaningful through tapping into social practices and the visitors' social context.

Keywords: Hybrid museum experiences; Meaning-making; Interpersonal experiences; New Museology

Virtual Reality. Augmented Reality. Artificial intelligence. Technological trends go through cycles of hype and bust, and in practice it is often more complicated to apply new technologies into real-world settings than the hype suggests. This is also the case in the museum domain, where the promises and challenges of digital technologies have permeated the field for decades. This is a major reason why this book approaches hybrid museum experiences as not relying on any particular technology – but as a design approach that is applicable to a range of technologies, making it possible to integrate digital media in physical contexts such as museum exhibitions. As such, the topic of this book is independent of technological trends.

In this chapter, we build on key insights from a wide range of academic disciplines to rethink what the museum experience *is* and *could be*. We outline shifts along each of the three dimensions identified in the introduction: In the physical versus digital dimension, a shift has occurred from viewing digital as a separate 'cyber' space to a view which focuses on how the digital integrates (or not) with our embodied, physical environments. In the personal versus social dimension, we point to a design potential in

---

Waern, A. and A. Sundnes Løvlie (eds.), *Hybrid Museum Experiences: Theory and Design*. Taylor & Francis Group, 2022
DOI 10.5117/9789463726443_CH02

addressing the *interpersonal*. And finally, in the museum versus visitor dimension we describe a shift in the field of museology, from focusing on museum collections to museum experiences (sometimes referred to as 'new museology').

The Physical and the Digital

In some of the first hypes around digital technology (e.g., in the 1980s and 1990s), popular images of technology often saw the development as a liberation from the constraints of the physical world. Digital technology was seen as creating a separate space, 'cyberspace', in which different rules applied. For instance, digital storage media were often described by estimating how many books that could fit into a small disk – eventually, entire libraries! The popular TV show *Star Trek* introduced TV audiences to the idea of the 'Holodeck', a room where all the walls were computer screens that could transport the user into a virtual world. In the 1990s, Virtual Reality (VR) first became commercially available, similarly inviting users to step out of the real world and into a virtual one. The rapidly increasing popularity of the World Wide Web introduced the broader public to the Internet, which gave the idea of a separate digital space – or universe – a radical new meaning: A global network which not only could store all the world's information, but which could be traversed in no time (depending on your dial-up modem connection).

However, a vision of a virtual, disembodied information space liberated from the constraints of physical space has a significant problem: That human beings have bodies, and our experiences of the world are very much rooted in experiences of physical space. There have long been countercurrents in computer research that focus not on cyberspace, but rather on embedding technologies in our real-world environment in order to augment and enrich it – e.g., through concepts such as ubiquitous computing or the Internet of Things.[1] Research in Human-Computer Interaction (HCI) has similarly suggested that technologies should be carefully integrated in our physical surroundings, and support our embodied experiences and meaning-making processes.[2] The popular breakthrough of smartphones

1 Weiser, 'The Computer for the 21st Century'; Mattern and Floerkemeier, 'From the Internet of Computers to the Internet of Things'.

2 Dourish, *Where the Action Is: The Foundations of Embodied Interaction*; Hornecker and Buur, 'Getting a Grip on Tangible Interaction'.

from the 2000s onwards has led to increased attention to the ways in which digital technology can integrate with our everyday environments. This has not only supported the commercial development of entertainment and services, but also facilitated new forms of art such as locative media and games such as pervasive games.[3] Some scholars suggest that the characteristics of smartphones – they are always-on and always-there, allowing us to carry with us a mobile computer connected to the global (and social) network – turn our everyday spaces into hybrid spaces that are 'blurring the borders between digital and physical spaces', characterising the technological development of the last decades as a trajectory 'from cyber to hybrid'.[4]

The use of digital technologies in museums mirror this broader development in some important ways. Many museum digitization efforts have focused on using the digitized collections to overcome limitations of space and time. For instance, making digitized collections available online help overcome spatial limitations in at least two ways: First, it becomes possible for people to see the museum's collections even if they are not able to visit the museum physically; and second, it also makes it possible for the museum to showcase the vast amount of their collections that are in storage and that they don't have room to exhibit physically. Perhaps the most prominent example of this is the European flagship project Europeana, imagined as an European digital library that gathered all of the continent's digitized cultural heritage in one information portal, available to people everywhere.[5] Similarly, the OpenGLAM movement argues that museums should work to make their digital collections as openly accessible as possible, to encourage creative reuse and 'engage audiences in novel ways on the web'.[6] However, the use of online collections seems to still be fairly moderate compared to physical visits to museums: For instance, Europeana's collections website attracted 6 million visits in the last year, which pales in comparison to the over 650 million visits to European museums per year, according to the most recent statistics from EGMUS.[7] A study of museum websites in Denmark showed that the websites were primarily used to access information used to plan a visit to the physical museum – such as opening hours and the

3 Montola, Stenros, and Waern, *Pervasive Games: Theory and Design*; Russell, 'Headmap Manifesto'; Tuters and Varnelis, 'Beyond Locative Media'.

4 De Souza e Silva, 'From Cyber to Hybrid: Mobile Technologies as Interfaces of Hybrid Spaces'.

5 Valtysson, 'Europeana: The Digital Construction of Europe's Collective Memory'.

6 https://openglam.org/principles/ (Accessed December 2020)

7 *Europeana Digital Service Infrastructure Annual report* (*DSI-4*); EGMUS – European Group on Museum Statistics, 'EGMUS – Statistics'.

physical address.[8] It seems that visiting a physical museum exhibition has a strong attraction that is hard for online museum collections to match. In other words, museums might do well to not only focus on how to use digital technology to make their digital collections accessible outside the museum – but also to pay close attention to how technology can be used to enrich experiences with the physical museum and its exhibitions. In other words, to explore hybrid technologies for museum experiences.

Museum Technology

Many types of technologies are used in museum spaces, including interactive information screens, audio/multimedia guides, interactive installations, VR headsets, and more. Hornecker and Ciolfi suggest that technology in museum spaces can be divided in three main interaction frames: Standalone installations, mobile interactions, and 'assemblies' that integrate digital technology 'across multiple spaces and exhibits'.[9] In stand-alone installations, technology is brought in as the dominant feature of a single exhibit with little connection to the physical collection. (A typical example is the digital information kiosk.) What is perhaps a more common way to bring technology into the museum is to add a layer of digital content to augment and complement the physical exhibition, using a device that the visitor carries. A well-established concept for such overlaid experiences are audio guides, originally offered on dedicated devices rented from the museum but now increasingly made available as mobile phone applications. More advanced examples than the audio guide include using virtual or augmented reality to let visitors see artefacts in new ways, such as seeing the Parthenon in Greece as it once was complete with colourful paintings, or seeing a statue as it was originally painted. The third format that Hornecker and Ciolfi discuss are assemblies, in which the physical exhibition and the technical augmentations (be they installations or mobile complements) 'work together as part of an overarching narrative or activity'.[10]

This book is primarily concerned with what Hornecker and Ciolfi call mobile interactions and assemblies. Both pose higher demands than stand-alone installations in terms of hybridity, in that the technology must be adapted to the physical constraints of the museum and its exhibits as well as

8 Holdgaard, 'Online Museum Practices. A Holistic Analysis of Danish Museums and their Users'.

9 Hornecker and Ciolfi, 'Human-Computer Interactions in Museums', pp. 18–32.

10 Hornecker and Ciolfi, 'Human-Computer Interactions in Museums', p. 29.

the (social) behaviour of visitors. Both forms also present rich opportunities for adding information, deepening perspectives, and providing alternative narratives.

This book is not about any particular technology solution. Many different technologies can be used in the design of hybrid museum experiences, including everyday technologies such as audio recordings and web apps, unusual and advanced technologies such as virtuality, or dedicated technology solutions for a particular museum. Examples from the GIFT project include smartphone web-apps (see Chapters 3 and 4), 3D-scanning (see Chapter 5), virtual reality, and even technologies for emotion sensing.[11] The case studies have been deliberately chosen to exemplify technologies that can be readily brought into most museums.

Common Risks Associated with Museum Technology

There are some challenges and risks that tend to emerge with the introduction of technology in the museum context, irrespective of which technology is chosen. These risks can be avoided through careful deliberation during the design and implementation of the technology, but it helps to be aware of them at the outset of any technology project.

Firstly, technology may steal focus from the physical exhibit. The most obvious way in which this happens is through stealing the visitors' perceptual attention; in particular, if the digital material is visual visitors may end up looking at screens instead of at the physical artefacts.[12] Since classical museums seldom involve listening to the artefacts on display, audio has long been the preferred way to convey digital content to audiences. Multiple ways to overcome this problem have been proposed in literature, but few are entirely successful. For example, Damala, et al. present a study of an information augmentation in which the visitor first had to physically direct the device towards the right object to obtain more information about it.[13] But

11 Back and others, 'GIFT: Hybrid Museum Experiences through Gifting and Play', MMCCXXXV; Darzentas and others, 'Card Mapper'; Løvlie and others, *Report on Sensitive Pictures*; Ryding, 'The Silent Conversation'; Spence, 'Using Hybrid Gifting to Build Personal Engagement in and with Museums'; Spence and others, 'Seeing with New Eyes: Designing for In-the-Wild Museum Gifting'; Spence and others, 'VRtefacts'; Tennent and others, 'Thresholds'.

12 Hsi, 'A Study of User Experiences Mediated by Nomadic Web Content in a Museum'; Petrelli and others, 'Integrating Material and Digital: A New Way for Cultural Heritage'; Tennent and others, 'Thresholds: Aligning Vision, Sound and Touch in Substitutional Reality'; Woodruff and others, 'Electronic Guidebooks and Visitor Attention'.

13 Damala and others, 'The Loupe'.

even so, though the activity involved actively searching for and locating the right object, the augmentation was seen as distracting by a large proportion of the study participants. It is worth noticing that even augmented reality devices that allow the user to 'see through' the screen, will put a layer between the visitor and the object that can make the experience feel less authentic.[14] Overcoming this problem requires that designers deliberate what is the desired balance in visitor attention, how much they should pay attention to the digital content versus to the museum, to then try to design their technology (and when possible, also the museum space and exhibition) to come as close as possible to this balance.

Another way in which hybrid museum experiences can steal focus is more insidious, and relates to how they offer opportunities for new museum experiences. For example, when discussing the opportunity of bringing games into the museum, Hornecker and Ciolfi point out that 'there is a risk with scavenger hunts and other similar gamified approaches to the visit that participants will only be concerned with the mechanics of the quest and concentrate on the competitive aspects, rather than on the heritage and linked narratives they relate to'.[15] The citation highlights how museums have inherent meanings and intended takeaways for museum visits, the 'heritage' and its 'linked narratives'. Unless these are carefully thought through and made explicit in the initial stages of a design project, they risk being overshadowed by the introduction of technology.

A second risk emerges from the way in which the museum offers a very different context for technology design than what most technology developers are used to. In a comparative investigation of over 108 exhibitions, Serrell found that visitors on average spend less than 20 minutes in an exhibition, independent of its size.[16] This also implies that visitors spend less time per singular exhibit in large exhibitions, than in smaller ones. Studies in art museums have shown that museum visitors spend surprisingly short time studying canonical works of art – typically around 15-30 seconds.[17] While this behaviour may be influenced by technology installations, it can't be expected to be fundamentally changed; hybrid museum installations must cater for short periods of engagement just as ordinary exhibits must. Even if a hybrid experience succeeds to engage some visitors for longer periods, the

14 Tolmie and others, 'Supporting Group Interactions in Museum Visiting'.

15 Hornecker and Ciolfi, 'Human-Computer Interactions in Museums', p. 27.

16 Serrell, 'Paying Attention'.

17 Smith, Smith, and Tinio, 'Time Spent Viewing Art and Reading Labels'.

design should not expect all visitors to do the same. Furthermore, a museum installation must often be designed so that visitors can start and end their engagement freely. A study of Flypad, a permanent multi-user augmented reality installation in an art gallery, found that visitors would 'rapidly and repeatedly disengage from and reengage with "Flypad"', including competing with each other for the controls and taking over each other's sessions with the installation.[18] The observation is typical for the ways in which visitors engage with interactive museum exhibits, and has multiple implications for their design. For example, if there are instructions for how to use the system that are shown only at the beginning of a session, visitors may not have seen those when they take over a session from another visitor. Another observation made frequently during GIFT, is that visitors tend to find it much easier to understand how interactive installations work when there is a large number of visitors present in the same room, as these give clues for how to interact.

These kinds of behaviours are rare outside the museum and science centre contexts, which is why even experienced technology designers may not foresee them. Common mistakes in the design of museum technology include designing for long periods of engagement, not supporting hand-overs, and designing interfaces that are either too private (so that visitors cannot learn by watching each other), or too public (e.g., so that visitors get exposed to the solution to a puzzle while still trying to solve it).

A third risk emerges from the ways in which information technology has a very long history of being designed for single users. Knowledge on designing technology for group interaction is both more recent, and tends to be less well known among technology designers. Museums tend to be visited in groups and no matter what technology is employed, the design has to take this into account. In the next subsection, we will dig into the design of social experience in more detail. Here, we will only note that this tends to create problems when using off-the-shelf technology in the museum context. A typical example of this is the use of VR helmets in museums, of which we saw multiple examples at the time when GIFT was initiated. All of these installations were severely limited in the number of visitors who could use the VR helmets at the same time, leading to queues, waiting times, and signup schedules that required additional personnel resources. In addition, the VR helmet would isolate one visitor from their group, leading to groups splitting up, or waiting idle, while one of the members were attending to the VR experience. Designing for VR in museums means that one must also

18 Flintham and others, 'Flypad'.

take these limitations into account, for instance by creating an interesting spectator experience for those that are waiting in line.[19]

The Personal and the Social

In recent years, much effort has been directed at creating personalized museum experiences: Experiences that are tailored to a specific person or group.[20] Museums have sometimes turned to technology as a way to offer personalized views of their collections, often in conjunction with giving access to additional information about the objects on display, or even to digitalized versions of objects in their collection that are not on display. The general idea of such solutions is that they will offer a richer and deeper insight into the museum collection, while still not overburdening the visitor with information that they are not interested in. Museums have focused on ways in which such rich information sources can be tailored to the individual visitor, both by filtering content to fit the individual, and by adapting the style of presentation.

However, there is a problem with this perspective: While the concept of personalization may present visitors with a higher level of agency than the traditional audio guide or guided tour, it does not offer a perspective on how visitors may engage in dialogue with the museum message, *nor with each other*. Visitors seldom come alone to museums.[21] Research rather shows that the visitor group is a social unit where group pressures are often given priority over individual preferences.[22] Blud even claims that 'interaction between visitors may be as important as interaction between the visitor and the exhibit', and the meaning-making processes that take place within the museum are to a large extent social.[23]

Museum exhibits and guides are often led by a one-way perspective on meaning-making: They supply information in a way that places the visitor in

19 Ioannidis, Eklund, and Løvlie, 'A Lifecycle Study of a Substitutional Reality Installation in a Museum Space'.

20 Bohnert and Zukerman, 'Non-intrusive Personalisation of the Museum Experience'; Not and Petrelli, 'Blending Customisation, Context-Awareness and Adaptivity for Personalised Tangible Interaction in Cultural Heritage'.

21 Hooper-Greenhill, 'Changing Values in the Art Museum'.

22 McManus, 'What People Say and How They Think in a Science Museum'.

23 Blud, 'Social Interaction and Learning Among Family Groups Visiting a Museum', p. 43; Eklund, 'A Shoe Is a Shoe Is a Shoe: Interpersonalization and Meaning-making in Museums – Research Findings and Design Implications'; Tolmie and others, 'Supporting Group Interactions in Museum Visiting'.

a role as a passive and isolated recipient of information. But meaning-making is also, and perhaps primarily, a social and shared process.[24] Shared experiences, narratives, and stories are central resources in meaning-making, and most classroom pedagogics are strongly reliant on shared meaning-making and dialogic processes.[25] We can tap into this understanding in order to create socially meaningful museum experiences.

Interpersonalization

A useful way to rethink the social aspects of a museum visit is through the lens of *interpersonalization*.[26] This concept captures processes of meaning-making that take place *between people* through their experience of a museum, rather than between the museum and an individual. Interpersonal meaning-making can happen during a visit, but also takes place before and after a visit, as people discuss their plans and experiences with friends, family, and colleagues. Your visit is likely to be influenced by other people, whether they are present in the flesh or not.

In sociology, the concept of *social ties* is used to understand how people are linked to each other.[27] Social ties can roughly be categorized as weaker or stronger, and they form through aspects such as how much time you spend together, how much you know about each other, and the practical things you do for each other (such as care and chores). Family bonds form the prime example of strong ties, but friendships can also be strong ties. By comparison, the tie to co-workers is typically weaker. Ties change over time; we might become close friends to co-worker, or break up with a romantic partner yet meet occasionally through mutual friends.

Museum visits are related to social ties in multiple ways. Firstly, a joint museum visit is a way to spend time together, and can be seen as something we do both *for* each other and *with* each other; hence they can serve to strengthen social ties. This must be taken into consideration when designing for groups of visitors: They come to the museum not only to experience the exhibition but also to do something together, and often the social aspect of the visit may be at least as important to them as the exhibition itself.

24 Drath and Palus, *Making Common Sense*.

25 Mortimer and Scott, *Meaning Making in Secondary Science Classrooms*; Rosen, 'The Importance of Story'; Short, 'Story as World Making'; Trevarthen, 'The Generation of Human Meaning'.

26 Eklund, 'A Shoe Is a Shoe Is a Shoe: Interpersonalization and Meaning-making in Museums – Research Findings and Design Implications'; Ryding and others, 'Interpersonalizing Intimate Museum Experiences'.

27 Granovetter, 'The Strength of Weak Ties'.

Managing these overlapping and occasionally contradictory goals is an important part of the visitor's experience and should be accounted for in design. Secondly, even when we do not visit the museum together with our close friends and family we will often want to share something of our experience with them. This desire to share the experience with our close social ties can be capitalized upon in the design of museum experiences. Furthermore, our close social ties to other people can also add to our own perspective by allowing us to understand their perspective on the museum – allowing us to 'see the museum through their eyes'.[28] This is what makes interpersonal experiences radically different from mere personalization.

Social ties are created from and through social actions.[29] Visiting a museum together is one example of an event which can consist of a series of social actions. Social actions are oriented towards other persons, and can be both synchronous and asynchronous; others can be individuals or groups, known or unknown, present or absent. Looking at social action and how these are part of creating and supporting social ties between people, we can identify social practices that stand out as particularly conducive for museum experiences. Hybrid museum experiences can be designed to draw upon close ties by asking visitors to engage in social actions, and through that enable visitors to form interpersonal experiences. Below, we will discuss more in depth practices related to *gifting* and *play*, that fruitfully can be tapped into in the design of hybrid museum experiences.

From Sharing towards Gifting

When we shift the perspective on meaning-making from the individual to the social, we open up the possibility of tapping into human social practices as vehicles for meaning-making, as they present forms of meaning-making that the visitors will recognize. One such practice is *gifting*. In Chapter 3, we present a case study of a hybrid experience capitalizing on gifting. Here, we introduce the theoretical underpinnings of gifting and its relevance for the museum context, in general and as an introduction to the case study to follow.

Gifting has long been studied in anthropology and is ubiquitous in human culture.[30] The common understanding of gifting is as a practice that serves to

28 Spence and others, 'Seeing with New Eyes: Designing for In-the-Wild Museum Gifting'.

29 Weber, 'The Nature of Social Action'.

30 Malinowski, *Argonauts of the Western Pacific: An Account of Native Enterprise and Adventure in the Archipelagoes of Melanesian New Guinea*; Mauss, *The Gift*.

strengthen, maintain, or at times weaken social ties.[31] Museums already tap into this practice in that many museums include a gift shop. Creating and giving a gift is personal and unique: It is something we usually only do for those we are closest to (or at the very least, close enough that we get invited to their birthday party). Designing to support gifting practices is a strong way to enable interpersonalization, because the act of selecting or creating a gift shifts the focus from the *self* to the *other*, from the person having the visit to the person the visitor wants to reach out to. The norms and conventions for gift-giving are fundamentally interpersonal, and suggest that a good gift is usually something that has been carefully chosen to fit the recipient. It is a unique fit to their specific personality, and perhaps it also refers to some aspect of the relation between giver and recipient, serving to reaffirm and strengthen their bond. Tapping into gifting practices also allows the museum to reach out to the visitors' social ties even when the recipients are not already visitors. Maybe they will come there later to experience the gift?

Gifting is fundamentally different from the sharing experiences enabled by social media such as Facebook, Twitter, or Instagram. Museums meet with a host of mobile-phone-based practices that centre around sharing. Visitors take photos in the museum and share them over social media, they check in to the museum, they may 'like' posts from the museum, etc. Some museums see this as a problem – picture-taking in particular is sometimes prohibited both for preservation and copyright reasons. However, many museums have begun to orient themselves towards these practices, which are seen as a way of reaching out, a 'viral marketing' channel. Gifting differs from social media sharing in that it is normally directed towards one specific individual (the receiver), rather than a larger audience (such as social media 'friends' or 'followers'), and thus implicates a much more personal or even intimate mode of communication.

The Museum and the Visitor

Designing for the personal and the social also entails that the museum's authoritative voice is given somewhat lower priority, in order to make room for the visitors' perspectives. This is in line with many ideals associated with both new media and the museological direction known as new museology, which often favour dialogical formats over one-way, monological communication. However, this change comes with some challenges. Museums and

31 Davies and others, 'Gifts and Gifting'.

cultural heritage sites are institutions with a strong cultural significance, and both professionals and visitors have strong ideas about their significance, including how to behave in the museum space.

Traditionally, museums have been seen as institutions responsible for conserving and classifying historical objects.[32] But let us look at what happens inside the standard museum. Some visitors may be walking alone, reading signs and contemplating the objects in view, adapting to what they perceive as the museum's canonical form of visit. But others do very different things. A visiting family may be rounding up their children, pointing at objects and reading signs, or just trying to locate the restroom. A romantic couple may be seeking out pictures of lovers, touching and smiling when they recognize their love as eternal. Other visitors may be rejecting, ridiculing, and reformulating the exhibits, laughing and pointing at something they find particularly ugly or quaint. Somebody snaps a selfie in front of a statue and publishes it on social media. As already discussed, all of these people are involved in *meaning-making processes*, and these are to a large extent social.

New Museology

Museums preserve, guard and shape a large part of our cultural heritage.[33] In one famous article, museums are even described as *temples* which enshrine 'The Works of God Through All the Ages; the Arts of Man Through All the Years'.[34] But increasingly, museums have started to be classified as leisure centres, turning the attention of institutions towards quality of user experiences focusing on joy and engagement.[35] The museum sector has increased its efforts in audience research, marketing, and customer profiling. A phenomenon referred to by some as 'primacy of the visitor' has begun to shape how museums organize themselves, considering the visitor a central point of focus when new plans are made.[36] In the extreme case, this has meant that the collection, the structure of space and all other elements of museum practice has been dominated by target group analysis, and the prediction of marketing departments have changed the focal point of museum activities. Curators have been steered towards designing for engagement and accessibility, and the prior focus on collections have

32 Greenhill, *Museums and the Shaping of Knowledge*.
33 Greenhill, *Museums and the Shaping of Knowledge*.
34 Cameron, 'The Museum, a Temple or the Forum', p. 17.
35 Middleton, 'The Future Demand for Museums 1990-2001'.
36 Macdonald, *A Companion to Museum Studies*, XXXIX.

shrunk into the background. This development has gone hand in hand with a rapid adoption of technology, as an element associated with fun and entertainment. Digital solutions such as displays, touch interfaces, games, and augmented and virtual reality have proliferated within the cultural heritage industry.[37]

A second and somewhat different development has contributed to this shift. In the late 80s and early 90s, political and social changes prompted a critical perspective on the practices of cultural heritage.[38] Many museum collections are aggregations of items from different parts of the world, some acquired through private donations, some through purchase while some are stolen goods, or war pillage. Even classification systems have been criticized as flawed, or even racist, in their focus on a historical narrative shaped by the white western world. With the rise of such post-colonial perspectives, many museums found themselves questioning their own identity.[39] One path towards reconciling the historical background of museums with being situated in a multi-cultural world, was to refocus on their local contexts and communities which (at least in cities) tend to be ethnically and politically pluralistic; and focusing on the contributions that museums could make to a modern society (IBID). Again, visitors and their perspectives, needs, and reasons to visit the museum, were brought into the limelight.

One of the ideas that emerged from this process was an ideal suggesting that museum visits were expected to be *transformative*; the visitor should emerge from the visit changed, having gone through a personal meaning-making journey. Both Bell and Soren describe this as a prevalent motivation among museum professionals for visitors to not only learn something, but to have a meaningful and deep experience that has an impact beyond the visit – that somehow will change, or transform, the visitor.[40] But designing for transformative experiences is in no way easy. The meaning that museum visitors create is dependent on all those things the visitors themselves bring with them: Their history, body, sense of identity, and previous experiences.[41] The visit is also so much more than their engagement with artefacts and digital content: Their physical surroundings, the physical sensations they

37 Witcomb, 'The Materiality of Virtual Technologies: A New Approach to Thinking about the Impact of Multimedia in Museums'.

38 Hooper-Greenhill, *Museums and the Shaping of Knowledge.*

39 Merriman, 'The Crisis of Representation in Archaeological Museums'.

40 Bell, 'Making Sense of Museums: The Museum as "Cultural Ecology"'; Soren, 'Museum Experiences that Change Visitors'.

41 Rodley, 'Playing With the Past, Part Two'; Silverman, 'Visitor Meaning-Making in Museums for a New Age'.

have, the pre-reflective emotions they experience, and the way they interact with other people in the museum will all influence their meaning-making process. This process begins before and ends long afterwards the physical visit. The meaning-making processes of museum audiences are thus only partly under the control of the museum.[42]

Today, museums are hybrid spaces. Visitors do a variety of things inside museums: They look at exhibitions, learn, enjoy time together with their family, take photos, drink coffee in the cafeteria, play with interactive displays, buy gifts in the shop, post on social media, send messages to an absent friend, and much more. Some of these activities are limited to the physical space and the time of the visit, such as discussing an artefact with fellow visitors; some reach outside of the experience, such as taking a selfie and posting it on social media. It is worth noting that even if we disregard those practices that rely on digital technologies, the museum space serves a multitude of activities and purposes for a variety of visitors and professionals, and thus may be considered a social and institutional hybrid even without regard to the complicating factor of technology.

From Storytelling towards Playful Meaning-Making

Playful museum experiences offer a path towards allowing visitors to engage with the content of the museum in ways that go beyond passive reception of museum messages. A key component of playful engagement is a process of resignification, through which participants collectively agree that artefacts, people, places and actions for a moment will take on a different meaning and role in the game.[43] Play offers rich opportunities for social interaction as human play is – almost always – a social activity.[44] This means that in play, the meaning of museum objects and spaces can be reinvented and the meaning-making process as such can be co-creatively explored.

With adults, play typically (but not always) takes the form of games with set rules and goals.[45] The use of rules and goals creates alibi for behaving differently – there are things you do in games that you would not do otherwise. One example is the game *Tombstone Hold 'Em*, designed to

42 Eklund, 'A Shoe Is a Shoe Is a Shoe: Interpersonalization and Meaning-making in Museums – Research Findings and Design Implications'.

43 Bateson, 'A Theory of Play and Fantasy'.

44 Huizinga, *Homo Ludens: A Study of the Play-Element in Culture.*

45 Caillois, *Man, Play and Games.*

be played in historic cemeteries partly as a way to increase the citizen engagement needed for heritage conservation – and partly to create a playful way to contemplate death.[46] Inspired by the popular poker variety *Texas Hold 'Em*, the goal of the game is to create a winning hand of tombstones, resignified to be playing cards within the game depending on the conventional symbols used on most of the tombstones. Players work in pairs to find the best pair of tombstones/cards that they can physically touch while holding hands. In order to achieve this goal, players have to explore the cemetery and study the tombstones, providing an encounter with the cultural heritage which at a minimum teaches them about the design conventions of the tombstones, and possibly also fosters an increased interest in the heritage and its conservation. Since the game is played in pairs, it capitalizes on pre-existing social ties and interpersonal meaning-making, providing players with a shared, social experience to remember.

However, games with challenges and winning conditions are not the only types of playful activities that can take place in a museum. In Chapter 4, we will present an example of a hybrid experience for adults which is designed to be a structured play activity but is very far from being a competitive game.

A common mistake in thinking about play is that it is a way to bring in fun and enjoyment. But play is not always perceived as 'fun'.[47] Play comes in many forms. One form of play that is particularly appropriate for the museum context is *pretense play*; visitors can, for example, pretend to be a historical person or to interact with somebody from another time period.[48] Museum activities for children will often include various forms of creative play, the playful creation of your own content. Getting adults to engage in creative play is harder as they tend to be more self-critical, but it can be done. Play activities can also focus on creating *immersive* experiences, encouraging visitors to enter a different state of mind through reflecting intensely their sensations or emotions. Finally, it can sometimes be useful to include ways for visitors to joke, create satire, or ridicule the museum. Satire creates a critical distance from the museum content, which for some visitors might be necessary in order to engage with it at all.

Games in the museum context have however not always been successful. A particular risk arises from *shallow gamification*, experiences that

46 McGonigal, *Reality Is Broken*.

47 Sicart, 'Toward an Ethics of Homo Ludens'.

48 Bretherton, 'Pretense: The Form and Function of Make-Believe Play'.

focus on introducing scores and rewards to foster motivation.[49] These very easily lead to participants focussing entirely on the rewards, rather than on the actual content of the museum. When shallow gamification fails, you will see the players actively trying to bypass looking at objects or reading texts, in order to access the next challenge in the game. They may for example just press the buttons in a quiz over and over again until they get the right answer by chance, rather than trying to figure out what is the right answer. This does not mean that shallow gamification is always bad, or that games in general only provide bad design options for museums. But it shows that playful experiences in the museum must be carefully designed, maintaining focus on what the designers want players to *do* and *experience*, rather than on the reward system or for what players should be rewarded.

Concerns with participation

In recent years, there has been an ongoing shift in attitudes towards participatory experiences. Early developments – such as the popularity of computer games, the invention of the web in the 1990s, and the so-called Web 2.0 in the 2000s – carried with them an optimism about participatory media, which were seen to enable more dialogic and inclusive experiences, giving 'ordinary people' more ways to make their voices heard. In the museum domain these ideals were captured in Nina Simon's influential book *The Participatory Museum*.[50]

However, in recent years there has been growing concern about some problematic aspects of participation, in some areas amounting to a backlash over social media and other forms of online participation. One quite widespread concern regards the fear that smartphones, social media and other communication technologies have replaced face-to-face sociability, thus impoverishing human relations.[51] In popular discourse these concerns are often linked to the idea that digital technologies are addictive, and many people attempt to undergo a 'digital detox'.[52]

In a separate but related development, the practice of facilitating commenting and debating online, which became widespread through the web 2.0

49 Deterding, 'Getting Gamification Right'.

50 Simon, *The Participatory Museum*.

51 See, e.g. Turkle, *Alone Together: Why We Expect More from Technology and Less from Each Other.*

52 Syvertsen and Enli, 'Digital Detox'.

technologies in the early 2000s, has increasingly been criticized for offering platforms for misogyny, hate speech, harassment, conspiracy theories, etc. Some prominent examples include the Gamergate phenomenon and debates connected with the #metoo movement.[53] Such concerns have lead many online media outlets to either remove online comments from their platforms, or to reduce their visibility, leading the influential *Wired* magazine in 2015 to declare 'the end of the comments'.[54] The scandal surrounding the firm Cambridge Analytica and their use of data from Facebook users led to concerns about the influence that social media might have on elections, and the debate around fake news and conspiracy theories has further fuelled a general distrust in social media. In 2020, the CEOs of both Facebook and Twitter were called to repeatedly defend their companies' practices in hearings before the US Senate.

All of these concerns demonstrate that it is far from risk-free for museums (or anyone else) to set up digital platforms that invite visitors and others to contribute with comments and other exchanges. One can no longer naïvely cite utopian ideals about the democratizing potential of digital technology, but must employ more nuanced and sophisticated approaches that carefully facilitate and scaffold participation in specific, focused ways that allow for positive contributions while guarding against the many disruptive pitfalls of participation.

One important benefit of hybrid experiences is that they focus interaction on the physical presence in the museum space, which both helps to create context for the interaction and limits some of the destructive potential in opening up 'the floodgates' to the wider online communities. Furthermore, designing for interpersonal experiences has great benefit in focusing on intimate experiences shared between two (or a few) people, which on the one hand shields the participants from having to relate to a larger online public, and furthermore may help reduce the negative consequences if participants choose to use the systems in destructive ways (e.g., an offensive joke will hopefully have much less negative impact if it is sent in a message to one of the participant's contacts, than if posted to a museum's website).

In the final chapter of this book, we will return to discuss more in depth some of concerns regarding how hybrid museum experiences may challenge and change the museum visit.

53 Chess and Shaw, 'A Conspiracy of Fishes, or, How We Learned to Stop Worrying About #GamerGate and Embrace Hegemonic Masculinity'.

54 Finley, 'A Brief History of the End of the Comments'.

Conclusions

Technology affords new and different practices that can take place in the museum. As such, it creates both challenges and opportunities for museums. In this chapter, we have explored these opportunities and challenges through the multifaceted lens of hybridity. This lens helps us see that solutions do not lie solely with the technology, but in the careful crafting of the meeting between technology, the physical museum, museum professionals, visitors and the social practices in museums. Successful design thus requires that the design team understands both the museum and what it intends to communicate, and the ways in which visitors will approach and make sense of the experience both socially and individually.

In subsequent chapters we will look at example designs that, in very different ways, offer deeply meaningful hybrid museum experiences.

Bibliography

Jon Back, Benjamin Bedwell, Steve Benford, Lina Eklund, Anders Sundnes Løvlie, William Preston, and others, 'GIFT: Hybrid Museum Experiences through Gifting and Play', in *Proceedings of Workshop on Cultural Informatics*, ed. by Angeliki Antoniou and Manolis Wallace (presented at the EUROMED2018, Nicosia, Cyprus: CEUR Workshop Proceedings, 2018), MMCCXXXV, pp. 31–40 <http://ceur-ws.org/Vol-2235/paper4.pdf> [accessed 3 June 2021]

Gregory Bateson, 'A Theory of Play and Fantasy', *Psychiatric Research Reports*, 2 (1955), 39–51

Genevieve Bell, 'Making Sense of Museums: The Museum as "Cultural Ecology"', *Intel Labs* (2002), 1–17

Linda M. Blud, 'Social Interaction and Learning Among Family Groups Visiting a Museum', *Museum Management and Curatorship*, 9.1 (1990), 43–51

Fabian Bohnert, and Ingrid Zukerman, 'Non-intrusive Personalisation of the Museum Experience', in *User Modeling, Adaptation, and Personalization*, ed. by Geert-Jan Houben, Gord McCalla, Fabio Pianesi, and Massimo Zancanaro, Lecture Notes in Computer Science (Berlin, Heidelberg: Springer, 2009), pp. 197–209 <https://doi.org/10.1007/978-3-642-02247-0_20>

Inge Bretherton, 'Pretense: The Form and Function of Make-Believe Play', *Developmental Review*, 9.4 (1989), 383–401 <https://psycnet.apa.org/doi/10.1016/0273-2297(89)90036-1>

Roger Caillois, *Man, Play and Games* (Champaign, IL: University of Illinois Press, 2001)

Duncan F. Cameron, 'The Museum, a Temple or the Forum', *Curator: The Museum Journal*, 14.1 (1971), 11–24 <https://doi.org/10.1111/j.2151-6952.1971.tb00416.x>

Shira Chess, and Adrienne Shaw, 'A Conspiracy of Fishes, or, How We Learned to Stop Worrying About #GamerGate and Embrace Hegemonic Masculinity', *Journal of Broadcasting & Electronic Media*, 59.1 (2015), 208–220 <https://doi.org/10.1080/08838151.2014.999917>

Areti Damala, Eva Hornecker, Merel J. van der Vaart, Dick Van Dijk, and Ian Ruthven, 'The Loupe: Tangible Augmented Reality for Learning to Look at Ancient Greek Art', *Mediterranean Archaeology and Archaeometry*, 16.5 (2016), 73–85 <https://doi.org/10.5281/zenodo.204970>

Dimitrios Darzentas, Raphael Velt, Richard Wetzel, Peter J. Craigon, Hanne G. Wagner, Lachlan D. Urquhart, and others, 'Card Mapper: Enabling Data-Driven Reflections on Ideation Cards', in *Proceedings of the 2019 CHI Conference on Human Factors in Computing Systems*, CHI '19 (New York, NY, USA: ACM, 2019), p. 571:1-571:15 <https://doi.org/10.1145/3290605.3300801>

Gary Davies, Susan Whelan, Anthony Foley, and Margaret Walsh, 'Gifts and Gifting', *International Journal of Management Reviews*, 12.4 (2010), 413–434 <https://doi.org/10.1111/j.1468-2370.2009.00271.x>

Sebastian Deterding, 'Getting Gamification Right' (unpublished Slideshare presented at the Google Tech Talks, Mountain View, California, 2011) <http://www.slideshare.net/dings/meaningful-play-getting-gamification-right> [accessed 5 October 2015]

Paul Dourish, *Where the Action Is: The Foundations of Embodied Interaction* (Cambridge, MA: The MIT Press, 2001)

Wilfred H. Drath, and Charles J. Palus, *Making Common Sense: Leadership as Meaning-making in a Community of Practice* (Greensboro, NC: Center for Creative Leadership, 1994)

EGMUS – European Group on Museum Statistics, 'EGMUS – Statistics' <https://www.egmus.eu/es/statistics/> [accessed 23 June 2021]

Lina Eklund, 'A Shoe Is a Shoe Is a Shoe: Interpersonalization and Meaning-making in Museums – Research Findings and Design Implications', *International Journal of Human-Computer Interaction*, 36.16 (2020), 1503–1513 <https://doi.org/10.1080/10447318.2020.1767982>

Europeana Digital Service Infrastructure Annual Report (DSI-4) (Luxembourg: Europeana, 25 September 2020), p. 48 <https://pro.europeana.eu/post/europeana-dsi-4-annual-report-2019-2020> [accessed 4 June 2021]

Klint Finley, 'A Brief History of the End of the Comments', *Wired*, 8 October 2015 <https://www.wired.com/2015/10/brief-history-of-the-demise-of-the-comments-timeline/> [accessed 5 October 2020]

Martin Flintham, Stuart Reeves, Patrick Brundell, Tony Glover, Steve Benford, Duncan Rowland, and others, 'Flypad: Designing Trajectories in a Large-Scale Permanent Augmented Reality Installation', in *ECSCW 2011: Proceedings of the 12th European Conference on Computer Supported Cooperative Work, 24-28 September 2011, Aarhus Denmark*, ed. by Susanne Bødker, Niels Olof Bouvin, Volker Wulf, Luigina Ciolfi, and Wayne Lutters (London: Springer, 2011), pp. 233–252 <https://doi.org/10.1007/978-0-85729-913-0_13>

Mark S. Granovetter, 'The Strength of Weak Ties', *American Journal of Sociology*, 78.6 (1973), 1360–1380 <https://doi.org/10.1086/225469>

Eileen Hooper Greenhill, *Museums and the Shaping of Knowledge* (London: Routledge, 1992)

Nanna Holdgaard, 'Online Museum Practices. A Holistic Analysis of Danish Museums and their Users' (unpublished PhD thesis, IT University of Copenhagen, 2014) <https://pure.itu.dk/portal/en/publications/online-museum-practices-a-holistic-analysis-of-danish-museums-and-their-users(13c275c8-616f-4203-91aa-e835049f5102).html> [accessed 4 June 2021]

Eilean Hooper-Greenhill, 'Changing Values in the Art Museum: Rethinking Communication and Learning', *International Journal of Heritage Studies*, 6.1 (2000), 9–31 <https://doi.org/10.1080/135272500363715>

Eilean Hooper-Greenhill, *Museums and the Shaping of Knowledge* (London & New York: Routledge, 1992)

Eva Hornecker, and Jacob Buur, 'Getting a Grip on Tangible Interaction: A Framework on Physical Space and Social Interaction', in *Proceedings of the SIGCHI Conference on Human Factors in Computing Systems*, CHI '06 (New York, NY, USA: ACM, 2006), pp. 437–446 <https://doi.org/10.1145/1124772.1124838>

Eva Hornecker, and Luigina Ciolfi, 'Human-Computer Interactions in Museums', *Synthesis Lectures on Human-Centered Informatics*, 12.2 (2019), i–171 <https://doi.org/10.2200/S00901ED1V01Y201902HCI042>

S. Hsi, 'A Study of User Experiences Mediated by Nomadic Web Content in a Museum', *Journal of Computer Assisted Learning*, 19.3 (2003), 308–319 <https://doi.org/10.1046/j.0266-4909.2003.jca_023.x>

Johan Huizinga, *Homo Ludens: A Study of the Play-element in Culture* (Angelico Press (Reprint 2016), 1955)

P. Ioannidis, L. Eklund, and Anders Sundnes Løvlie, 'A Lifecycle Study of a Substitutional Reality Installation in a Museum Space', *Journal on Computing and Cultural Heritage*, Forthcoming (2020)

Anders Sundnes Løvlie, Edgar Bodiaj, Karin Ryding, Steve Benford, Harriet Cameron, Dimitrios Darzentas, and others, *Report on Sensitive Pictures*, The GIFT Project (Brussels, Belgium: The Community Research and Development Information Service (CORDIS), The European Commission, 20 January 2020), p. 52 <https://

ec.europa.eu/research/participants/documents/downloadPublic?documentIds=080166e5cb4633ce&appId=PPGMS> [accessed 4 June 2021]

Sharon Macdonald, *A Companion to Museum Studies* (John Wiley & Sons, 2011), xxxix

Bronislaw Malinowski, *Argonauts of the Western Pacific: An Account of Native Enterprise and Adventure in the Archipelagoes of Melanesian New Guinea* (London: Routledge (reprint 1978), 1922)

Friedemann Mattern, and Christian Floerkemeier, 'From the Internet of Computers to the Internet of Things', in *From Active Data Management to Event-Based Systems and More: Papers in Honor of Alejandro Buchmann on the Occasion of His 60th Birthday*, ed. by Kai Sachs, Ilia Petrov, and Pablo Guerrero, Lecture Notes in Computer Science (Berlin and Heidelberg: Springer, 2010), pp. 242–259 <https://doi.org/10.1007/978-3-642-17226-7_15>

Marcel Mauss, *The Gift: Forms and Functions of Exchange in Archaic Societies* (London: W. W. Norton & Company (reprint 2000), 1925)

Jane Evelyn McGonigal, *Reality Is Broken: Why Games Make Us Better and How They Can Change the World* (Penguin Press HC, The, 2010)

P. McManus, 'What People Say and How They Think in a Science Museum', in *Heritage Interpretation, Vol 2: The Visitor Experience*, ed. by D.L. Uzzell (London and New York: Bellhaven Press, 1989), pp. 156–165

Nick Merriman, 'The Crisis of Representation in Archaeological Museums', in *Cultural Resource Management in Contemporary Society*, ed. by Hatton and MacManamom (Routledge, 2000), pp. 316–325

Victor T.C. Middleton, 'The Future Demand for Museums 1990-2001', in *The Museums Profession: Internal and External Relations*, ed. by Gaynor Kavanagh, Leicester Museum Studies Series (Leicester, UK: Leicester University Press, 1991), pp. 139–160 <https://www.abebooks.co.uk/Museums-Profession-Internal-External-Relations-Leicester/22671512755/bd> [accessed 29 October 2019]

Markus Montola, Jaakko Stenros, and Annika Waern, *Pervasive Games: Theory and Design* (Boca Raton, FL: CRC Press, 2009)

Eduardo Mortimer, and Philip Scott, *Meaning Making in Secondary Science Classrooms*, 1st edn. (USA: McGraw-Hill, Inc., 2003)

Elena Not, and Daniela Petrelli, 'Blending Customisation, Context-Awareness and Adaptivity for Personalised Tangible Interaction in Cultural Heritage', *International Journal of Human-Computer Studies*, Advanced User Interfaces for Cultural Heritage, 114 (2018), 3–19 <https://doi.org/10.1016/j.ijhcs.2018.01.001>

Daniela Petrelli, Luigina Ciolfi, Dick van Dijk, Eva Hornecker, Elena Not, and Albrecht Schmidt, 'Integrating Material and Digital: A New Way for Cultural Heritage', *Interactions*, 20.4 (2013), 58–63 <https://doi.org/10.1145/2486227.2486239>

Ed Rodley, 'Playing With the Past, Part Two: Magic Circles and Interaction Alibis', *Thinking about Museums*, 2018 <https://thinkingaboutmuseums.com/2018/03/12/playing-with-the-past-part-two-magic-circles-and-interaction-alibis/> [accessed 16 June 2021]

Harold Rosen, 'The Importance of Story', *Language Arts*, 63.3 (1986), 226–237 <https://www.jstor.org/stable/41405427> [accessed 16 June 2021]

Ben Russell, 'Headmap Manifesto' (headmap.org, 1999) <http://technoccult.net/wp-content/uploads/library/headmap-manifesto.pdf> [accessed 2 November 2009]

Karin Ryding, 'The Silent Conversation: Designing for Introspection and Social Play in Art Museums', in *Proceedings of the 2020 CHI Conference on Human Factors in Computing Systems*, CHI '20 (Honolulu, HI, USA: Association for Computing Machinery, 2020), pp. 1–10 <https://doi.org/10.1145/3313831.3376357>

Karin Ryding, Jocelyn Spence, Anders Sundnes Løvlie, and Steve Benford, 'Interpersonalizing Intimate Museum Experiences', *International Journal of Human–Computer Interaction*, 2021, 1–22 <https://doi.org/10.1080/10447318.2020.1870829>

Beverly Serrell, 'Paying Attention: The Duration and Allocation of Visitors' Time in Museum Exhibitions', *Curator: The Museum Journal*, 40.2 (1997), 108–125 <https://doi.org/10.1111/j.2151-6952.1997.tb01292.x>

Kathy G. Short, 'Story as World Making', *Language Arts*, 90.1 (2012), 9–17

Miguel Sicart, 'Toward an Ethics of Homo Ludens', in *Ludics: Play as Humanistic Inquiry*, ed. by Vassiliki Rapti and Eric Gordon (Singapore: Springer, 2021), pp. 21–45 <https://doi.org/10.1007/978-981-15-7435-1_2>

Lois H. Silverman, 'Visitor Meaning-Making in Museums for a New Age', *Curator: The Museum Journal*, 38.3 (1995), 161–170 <https://doi.org/10.1111/j.2151-6952.1995.tb01052.x>

Nina Simon, *The Participatory Museum* (Santa Cruz, California: Museum 2.0, 2010) <http://www.participatorymuseum.org/> [accessed 27 January 2016]

Lisa F. Smith, Jeffrey K. Smith, and Pablo P. L. Tinio, 'Time Spent Viewing Art and Reading Labels', *Psychology of Aesthetics Creativity and the Arts*, 11.1 (2017), 77–85 <https://psycnet.apa.org/doi/10.1037/aca0000049>

Barbara J. Soren, 'Museum Experiences that Change Visitors', *Museum Management and Curatorship*, 24.3 (2009), 233–251 <https://doi.org/10.1080/09647770903073060>

Adriana de Souza e Silva, 'From Cyber to Hybrid: Mobile Technologies as Interfaces of Hybrid Spaces', *Space and Culture*, 9.3 (2006), 261–278 <https://doi.org/10.1177/1206331206289022>

Jocelyn Spence, 'Using Hybrid Gifting to Build Personal Engagement in and with Museums' (presented at the Museums Matter: Interpreting, Sharing and Collaborating in Research, Nottingham, UK, 2019) <https://www.nottingham.

ac.uk/home/featureevents/2019/museums-matter-%E2%80%94-interpreting-sharing-and-collaborating-in-research%E2%80%94.aspx>

Jocelyn Spence, Benjamin Bedwell, Michelle Coleman, Steve Benford, Boriana N. Koleva, Matt Adams, and others, 'Seeing with New Eyes: Designing for In-the-Wild Museum Gifting', in *Proceedings of ACM CHI Conference on Human Factors in Computing Systems (CHI 2019)* (presented at the CHI 2019, Glasgow, Scotland, UK: ACM, 2019) <https://doi.org/10.1145/3290605.3300235>

Jocelyn Spence, Dimitrios Paris Darzentas, Yitong Huang, Harriet R. Cameron, Eleanor Beestin, and Steve Benford, 'VRtefacts: Performative Substitutional Reality with Museum Objects', in *Proceedings of the 2020 ACM Designing Interactive Systems Conference*, DIS '20 (New York, NY, USA: Association for Computing Machinery, 2020), pp. 627–640 <https://doi.org/10.1145/3357236.3395459>

Trine Syvertsen, and Gunn Enli, 'Digital Detox: Media Resistance and the Promise of Authenticity', *Convergence*, 26.5–6 (2020), 1269–1283 <https://doi.org/10.1177/1354856519847325>

Paul Tennent, Sarah Martindale, Steve Benford, Dimitrios Darzentas, Pat Brundell, and Mat Collishaw, 'Thresholds: Embedding Virtual Reality in the Museum', *Journal on Computing and Cultural Heritage*, 13.2 (2020), 12:1-12:35 <https://doi.org/10.1145/3369394>

Paul Tennent, Sarah Martindale, Steve Benford, Dimitrios Darzentas, Pat Brundell, and Matt Collishaw, 'Thresholds: Aligning Vision, Sound and Touch in Substitutional Reality', *The ACM Journal on Computing and Cultural Heritage (JOCCH)*, Forthcoming (2020)

Peter Tolmie, Steve Benford, Chris Greenhalgh, Tom Rodden, and Stuart Reeves, 'Supporting Group Interactions in Museum Visiting', in *Proceedings of the 17th ACM Conference on Computer Supported Cooperative Work & Social Computing*, CSCW '14 (New York, NY, USA: ACM, 2014), pp. 1049–1059 <https://doi.org/10.1145/2531602.2531619>

Colwyn Trevarthen, 'The Generation of Human Meaning: How Shared Experience Grows in Infancy', in *Joint Attention: New Developments in Psychology, Philosophy of Mind, and Social Neuroscience*, ed. by A. Seemann (Cambridge, MA, US: MIT Press, 2011), pp. 73–113

Sherry Turkle, *Alone Together: Why We Expect More from Technology and Less from Each Other* (New York: Basic Books, 2011) <https://www.basicbooks.com/titles/sherry-turkle/alone-together/9780465093656/> [accessed 23 June 2021]

Marc Tuters, and Kazys Varnelis, 'Beyond Locative Media: Giving Shape to the Internet of Things', *Leonardo*, 39.4 (2006), 357–363 <https://doi.org/10.1162/leon.2006.39.4.357>

Bjarki Valtysson, 'Europeana: The Digital Construction of Europe's Collective Memory', *Information, Communication & Society*, 15.2 (2012), 151–170 <https://doi.org/10.1080/1369118X.2011.586433>

Max Weber, 'The Nature of Social Action', in *Max Weber: Selections in Translation*, ed. by W. G. Runciman, trans. by E. Matthews (Cambridge: Cambridge University Press, 1978), pp. 7–32 <https://doi.org/10.1017/CBO9780511810831.005>

Mark Weiser, 'The Computer for the 21st Century', *SIGMOBILE Mob. Comput. Commun. Rev.*, 3.3 (1999), 3–11 <https://doi.org/10.1145/329124.329126>

Andrea Witcomb, 'The Materiality of Virtual Technologies: A New Approach to Thinking about the Impact of Multimedia in Museums', in *Theorizing Digital Cultural Heritage: A Critical Discourse*, ed. by Fiona Cameron and Sarah Kenderdine (Cambridge, MA: The MIT Press, 2007), pp. 35–48 <https://mitpress.universitypressscholarship.com/view/10.7551/mitpress/9780262033534.001.0001/upso-9780262033534-chapter-3> [accessed 23 June 2021]

Allison Woodruff, Paul Aoki, Amy Hurst, and Margaret Szymanski, 'Electronic Guidebooks and Visitor Attention', in *International Cultural Heritage Informatics Meeting: Proceedings from ichim01*, ed. by D. Bearman and F. Garzotto (presented at the International Cultural Heritage Informatics Meeting, Milano, Italy: Archives & Museum Informatics, 2001) <http://www.archimuse.com/publishing/ichim01_vol1/woodruff.pdf>

About the Authors

Anders Sundnes Løvlie does research on the intersection of design research and media studies, focusing in particular on experience design, locative media and play. Anders was the coordinator for the GIFT project, and has been involved in a number of design projects involving museums like the Munch Museum, Danish Architecture Center, the National Gallery of Denmark, Brighton Museum and the Frederiksberg Museums. Author of the book *The Rhetoric of Persuasive Games: Freedom and Discipline in America's Army*. Leads the research group on Media, Art and Design (MAD).

Steve Benford is the Dunford professor of Computer Science at the Mixed Reality Laboratory at the University of Nottingham where he directs the Horizon 'My Life in Data' Centre for Doctoral Training and the 'Smart Products' beacon. He previously held an EPSRC Dream Fellowship, has been a Visiting Professor at the BBC and was elected to the CHI Academy in 2012. He has collaborated with many artists over the past thirty years to create, tour and study interactive performances and installations with a view to gaining new insights into how humans can experience computers.

Lina Eklund is a lecturer in HCI at the Department of Informatics and Media at Uppsala University, Sweden. She has a PhD in sociology and is interested in social behaviour in and with designed, digital technologies. She has been a researcher at The Swedish National Museum of Science and Technology, where she worked with the exhibition of digital games. The work resulted in the *Play Beyond Play* exhibit. Current work focuses on uses and practises of digital technologies in museums, such as digital games and hybrid experiences.

Paulina Rajkowska is a PhD Candidate at the Department of Informatics and Media at Uppsala University. She comes from a mixed background of sociology, media studies and HCI. Her prior work includes writing on social phenomena in video games as well as application of critical theory to design practices. In her ongoing research she is looking at reflexive practice, playfulness and how both of those can be applied in projects, to improve the quality of their outcomes.

Jocelyn Spence is an interaction design researcher with an interest in performative experience design, performance, gifting, and cultural heritage; she is a Research Fellow in the Mixed Reality Lab at the University of Nottingham.

Annika Waern is a 'research through design' academic who has been researching technology-supported physical play and games for about fifteen years. Currently, she is conducting research in the areas of hybrid play in museums, children's play in outdoor settings, and circus training to foster proprioceptic skills. Waern has authored over 100 academic articles and book chapters, and been editor of 2 books and several journal special issues. She is a DiGRA (Digital Games Research Association) distinguished scholar and a HEVGA (Higher Education Video Game Alliance) fellow.

Cases

3. The *Gift* App – Gifting Museum Experiences

Jocelyn Spence

Abstract

This chapter presents the experience of designing, deploying, and playing an artistically designed web app that invites museum visitors to turn their visit into a 'gift' – a hybrid experience made of photos and audio recordings that they send to someone they love; as if they were making a mixtape, only with objects from a museum. We discuss this from the perspectives of the designers, museum professionals, and visitors, to explore how a fundamental social activity such as gifting can create new layers of meaning for museum experiences.

Keywords: Artist-led research; Research in the wild; Voice; Personal messaging; Context; Connection; Gifting

If one principle of a design for supporting museum visitors is to take advantage of commonly used technologies that will be familiar to most visitors and reduce the need for costly overheads for museum professionals, how can that design be something besides a visual distraction or just another type of audio guide?

This was the first conundrum that the GIFT project had to face. In a sense, one strand of this research resolved it by avoiding it entirely. This strand, led by the artist group Blast Theory and researchers from the University of Nottingham, did not even attempt to convey factual information to visitors. It did not attempt to gamify the museum experience. It did not try to direct visitors to longer, deeper, or more informed encounters with individual objects or exhibits. Instead, the design that emerged went straight to the heart of personalization through interpersonalization. It also addressed the inherent sociality of many, arguably most or all, museum visits. And it

Waern, A. and A. Sundnes Løvlie (eds.), *Hybrid Museum Experiences: Theory and Design.* Taylor & Francis Group, 2022
DOI 10.5117/9789463726443_CH03

did both of these things by getting its users to do two things: First to think of a person who may have nothing to do with the museum at all, and then to return their focus to the museum, heightened and intensified by the intention to consider what this person might like.

This simple approach cannot be thought of in terms of personalization as it has been conducted in the past, because no pre-visit questionnaire, embodied location information, or dynamic user modelling can ever hope to know what is in a visitor's mind and heart that day. But use a technology to ask a visitor to think of someone they would like to give a gift to, and the museum has directly accessed that visitor's personal life in a way their questionnaire-wielding staff or most expensive adaptive modelling tool never could. One visitor has been missing their mother recently, another has an upcoming date that a quirky, no-cost gift would be perfect for, and one wants to trade gifts with her sister, who's standing right next to her. Keeping another person in mind while visiting is indeed a layer between the visitor and the museum, but it is a layer that can act as a lens to shift and tighten their focus.

Asking visitors to think of another person also supports a wide range of social encounters, including those that cannot exist without some form of technological intervention. The visitor who misses her mother may focus on objects that make him feel closer to her and to home. The woman about to go on her second date may try to anticipate what her date will like while finding objects that reveal details about her own preferences. And the pair of sisters know everything about each other already and simply enjoy a new way of interacting with the museum and with each other – splitting up to keep their gifts secret from each other until they swap over and receive something delightfully unexpected. Parents take turns watching the children while making for each other or take the children to make something for mummy. Groups of teenagers work together to hunt for the things that excite them. Foreign exchange students send home a slice of their life that day. An elderly pensioner on his own is thrilled to be able to give a meaningful gift to family members who ordinarily will not allow him to spend any money on them. Any and all combinations of social experience are supported (and each example here is one we witnessed).

The *Gift* app is a smartphone- or tablet-based experience that encourages a museum visitor to create a digital gift for another person, and lets the recipient receive this gift. The app was developed by artist collective Blast Theory and has been trialled formally and informally on numerous occasions, from the spring of 2017 until the end of 2019. The app is a good example of how a hybrid museum experience can build on human social practices to facilitate the creation of personally meaningful connections between friends over museum objects.

In this chapter, we look at how the *Gift* app is experienced, the motivations behind its design, and why gifting provides a path towards personally meaningful museum experiences.

The Experience

'You're going to make a gift for someone special', says the intimate female voice in your headphones. 'They might be next to you now; they might be on the other side of the world.'

You are standing in the main hall of Brighton Museum and Art Gallery in Brighton, a seaside town south of London. The hall is full of large and small objects on display, a somewhat eclectic collection of art and design from the 20th century. In your hands you have your smartphone, running an app made by the artist group Blast Theory – and a voice in your headphones instructing you to see the museum through the eyes of a person you want to make a gift for.

> Let that person be your filter as you walk. Ignore the things you know they're not going to care about. Go and learn more about the objects you know nothing about. Keep walking as you start to browse slowly. The first thing you choose: Why is it right for this person? Is it going to trigger a memory of the time you have spent together? Does it shine out in their favourite colour? What was the thought – and why does it count?

Eventually, you find an object you would like to include in your gift. You take a picture of it, and the woman in your ear tells you to record a message. Maybe you feel a bit awkward, standing in an exhibition room with other visitors walking by, so you find an empty corridor where you can speak into the phone.

The voice in your ear asks if you want to find one or two more objects and record messages for them as well. You find a great second object, and this time it's not so awkward to speak out loud in the museum. Then you notice something next to it you never would have paid attention to before, and decide to send it to your friend, too.

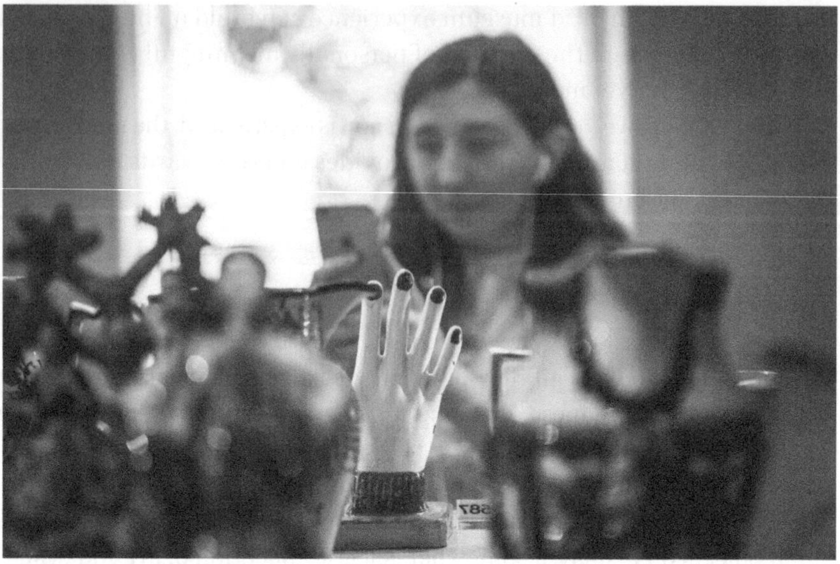

Figure 3.1. Visitor about to speak a message into the *Gift* app at Brighton Museum and Art Gallery, 2018. Photo by Jocelyn Spence.

The Designer Perspective

Interviews with Blast Theory artists Matt Adams, Nick Tandavanitj, and John Hunter reveal how the *Gift* experience developed from an originally very simple concept, made rich by a range of subtly nuanced design choices. Gifting seemed an ideal concept to build an app around, because gifting is a universal concept with nearly a century's research devoted to it, beginning with studies of gifting among Pacific Islanders and others.[1] In contemporary commercial cultures it has the potential to prompt emotions, shape relationships, and form important memories.[2] The artists aimed for simplicity, primarily in response to the needs articulated by museums including Brighton Museum and Art Gallery, to implement digital interventions that would enhance visitor engagement without demanding too much from the museum's financial or staffing resources. The goal was to create a self-contained, straightforward app that could be used with little up-front investment and minimal ongoing support.

1 Malinowski, *Argonauts of the Western Pacific: An Account of Native Enterprise and Adventure in the Archipelagoes of Melanesian New Guinea*; Mauss, *The Gift*.

2 Cheal, *The Gift Economy*; Ruth, Otnes, and Brunel, 'Gift Receipt and the Reformulation of Interpersonal Relationships'; Schwartz, 'The Social Psychology of the Gift'.

Rather than providing information, the app was to be an artistic experience. But why would a visitor use an app, even a simple one, if it wasn't obvious what it did, like an audio guide? How could the app be meaningful if it didn't give factual insights into the collection? The artists decided to build their app around a single proposal: To create a gift for a friend (anyone the visitor knows personally, encompassing partners, family members, close colleagues, and actual friends).

The designers wanted their app to reframe rather than compete with collections. Thus, they created a primarily voice-driven experience that would keep visitors' eyes on the collection and allow their attention to freely move between the collection, the objects on display, and their thoughts about the friend for whom they were making the gift. The tone and wording of the voice they would hear needed to convey the instructions for how to make a gift. It also needed to create a frame of mind that would encourage visitors to make an investment in a personal relationship. They aimed for an 'intensity' (quoting John Hunter) that would focus the visitor's attention on their relationship with their friend. They arrived at a voice that was aptly described as an 'intimate stranger' by a participant in the studies. They added to this intensity by requiring visitors to use their own voices to record short messages for their friend of why they had chosen each gifted object.

The app was developed in two main iterations. The first version ready for public release was launched during an intensive three-day deployment in July 2018 (Figure 3.1). Blast Theory staff and university researchers actively solicited participation and feedback from visitors who otherwise might not have tried the app. Details on this trial are reported in a recent study.[3] Significant effort was put into redesigning the app in light of this feedback and some notable changes were made, but the fundamental principles and strategies remained the same. Participant comments and observations in this chapter are generally drawn from the 2018 version but pertain to elements that remained consistent across the two iterations.

The main challenge for the second iteration, in July 2019, was to make a powerful experience easy to use when it was fully 'in the wild' with no human intervention beyond a minimal degree of promotion by the museum's own front of house staff. This posed a great challenge for Blast Theory, who primarily create artist-led experiences. The most obvious first step was to remove initial barriers to entry as far as possible by making Gift a web app requiring only a browser on a moderately up-to-date smartphone, with no downloads required. However, simply making it easier to access the app

3 Spence and others, 'Seeing with New Eyes: Designing for In-the-Wild Museum Gifting'.

would not be enough to meet the aims of the *Gift* app. In an interview after the second design iteration, artist Nick Tandavanitj commented upon this:

> Our experience is that one of the best ways to elicit seriousness, the kind of sense of deep engagement, is actually having a moment face-to-face with a performer who sets the tone, who sets a level of expectation and a level of framing as to what you're about to embark on [...] what is it [now] that we're giving you? [...] There's a tone that's set, and there's the level of familiarity and the language, that implies they know you and have a kind of relationship with you even though it's non-personal; and it's suggestive and it leads you through a process of thinking which is intended to be guiding you into a much more reflective space.

When there is no opportunity for face-to-face interaction, just face-to-screen – especially when people would rather be face-to-museum-collection than face-to-screen – audio becomes extremely important. In addition, Tandavanitj explains, the audio in the *Gift* app has to do much more than a typical audio guide ever would. It has to set the tone for an experience that most people have never done or even imagined: Giving a digital gift of a physical museum object. The tone and content of the narration have to accommodate people who methodically look for the perfect gift or those who roam in search of something that catches their eye; ones who struggle to create a masterpiece or ones who quickly send off a light-hearted 'stocking filler'. It has to communicate that giving a digital gift in a museum can be a playful activity at the same time as it might generate the kind of memorable connection that the best physical gifts do. Tandavantitj refers to this as the narration being able to 'acknowledge those different levels of subjectivity', and this 'balancing act' is Blast Theory's 'design challenge':

> While the first iteration was focused on the giving of gifts, in the second iteration a higher focus was placed on the receiver experience. Although the original design intended for both the giving and the receiving of gifts to happen in the museum, Blast Theory realised that most gifts would be sent to friends outside of the museum and sometimes very far away from it. This meant that the receiver experience could be almost entirely digital.

With the *Gift* app, receivers find out about their gift via email, SMS, WhatsApp, or Messenger. The receiving process takes receivers through the gift made for them in the sensible order: they can go to the museum, launch the web app from their phone, put on their headphones if they want to, hear the voiceover, follow the clue, see the photo, find the gift, and listen to their giver's audio message. Receivers who cannot or choose not to go to the museum can still get their gift wherever they are via the web app. Gifted objects might remind them of shared jokes or experiences, pique their curiosity about topics that the giver wanted to expose them to, or just let them know that the giver was thinking of them. The receiver can record a message for their giver at the end of the gift, if they wish. Tandavanitj provides Blast Theory's rationale for the design of the receiver's experience:

> When we set the palette of what the channels were, the primacy was to say is this a channel where you would expect to receive a message that's intended for you and is of value, it's not just a post in a Facebook stream, it's not just a post in an Instagram stream. It is actually something that I'm supposed to read and that was meant for me.

In other words, the *Gift* app was designed to structure different elements of the most commonplace of social media functions into a personally meaningful gift. That might mean a quick, light-hearted joke for some people, and a lengthy, effortful revelation for others, or anything in between. It might be welcome or, as with some physical gifts, unwelcome. But it would unquestionably be made for that one person, though made and delivered with the simplest of tools available to smartphone users.

The Museum Perspective

The *Gift* app was developed in close collaboration with Brighton Museum and Art Gallery, a perfect example of what Blast Theory's Tandavanitj refers to as 'organizations that really want to try things.' It is a local museum typical of its type and time in its mixture of collection types: The main hall's 20th-century art and design, a large collection of eighteenth- and nineteenth-century ceramics donated by the museum's founder, and sections devoted to Egyptology, images and artefacts charting the city's development, archaeology, fine art, fashion, trans identities (reflecting the city's proud reputation as the bastion of the UK's LGBT+ community), locally produced youth projects, temporary exhibitions, and more. The visitor populations are likewise a mixture of the local, with families making regular visits down the generations, and the global, given Brighton's popularity as an international tourist and student destination.

Blast Theory worked in close collaboration with Brighton Museum and Art Gallery throughout the development process. They met at various points with the digital team, staff members, and volunteers to solicit their expertise in communicating with each other and the public. The key to their collaboration was the museum's Digital Manager, Kevin Bacon, who offered his insights and assistance at every stage. Here he describes his vision of museum visits:

> The museum visit is a social one, and actually often the attention you have towards the cultural works is quite oblique, it's quite partial, and that's a real complexity to add to what the visitor experience is, especially if you're looking at designing in gallery elements to add to that. So in my experience, developing digital experiences or products for using galleries, it's often thinking about people who are only going to be half looking at something.

In this sense, Brighton provided a wonderful opportunity for a collaboration with an artist group 'having fresh eyes coming into the museum' and working 'within Brighton Museum as it is', including their concerns around conservation and visitor flow. Many visitor groups used the *Gift* app, from couples to small groups of friends to entire classes. From this museum's perspective, the app becomes another element of a social experience that acknowledges the complexity of visitor engagement and provides a new way of encountering objects in the collection without compounding issues around conservation or visitor flow.

The *Gift* app's use of social media is also very much in line with Brighton Museum's perspective, as well. As Bacon puts it, current use of social media as exemplified by the selfie tends to cast everything else, including a museum and its collection, as a backdrop to the individual. However, 'if we're thinking about the value of the museum and actually having a mechanism and a framework for people to share in a more structured, richer, more engaging and inspirational way, then I think yes, I saw *Gift* as being a platform to do that'.

It was important for museum staff to have at least a basic level of understanding and willingness to communicate the proposition of the app to visitors who otherwise might only see – or walk straight past – a poster or postcard. Much to Blast Theory's surprise, even the most engaged staff members actively invited relatively few visitors to take part. Often this was for understandable reasons: For example, an adult with three small children would likely view such an invitation as a source of unwanted pressure. However, many visitors with no obvious barriers to uptake were not approached. Bacon acknowledged 'the value of that very engaging front-of-house presence as a way of onboarding people onto the experience'. This of course requires staff to be 'able to take the time' and 'engage [visitors] in an experience whereby they would not only have a go at the app, but also that they're encouraged to come back and reflect on it'. While this can be a burden on staff members who already have plenty of tasks to manage, Bacon 'can see what a huge difference that makes to that sort of experience'.

Given the museum staff's positive attitudes and similar values, Blast Theory realised that they had stumbled onto an important lesson. The brief invitations to use the app that Blast Theory had assumed would be easy for staff and volunteers to incorporate into their routines proved difficult, sometimes insurmountable, from the points of view of some of these diligent, positive, and well-briefed museum workers. Blast Theory were also surprised to see visitors walk straight past large, colourful banners and a multitude of beautifully designed postcards inviting engagement. These phenomena underscore the need to design an app *in the context of the museum*, including the staff's relationship with the app, and visitors' expectations and priorities.

The Visitor Perspective

This analysis of the visitor experience is based primarily on our three-day test of the *Gift* app at Brighton Museum and Art Gallery in July 2018. Over those three days, researchers from across the project consortium and Blast Theory staff conducted 57 in-person interviews and received a further 57 short questionnaires.

The app affected an individual's experience in terms of how they *saw* the museum. The most frequent response when asked how to describe their experience was to speak in terms of seeing 'through fresh eyes', 'with other eyes', or similar. Some attributed their reaction to the explicit instruction to look for something their receiver would like. In this case, the difference emerged from their attempt to see things the way their receiver would see them. Others simply saw the app's instructions as a way to step out of their own expectations of what they 'should' find interesting.

The app also affected many visitors' ways of *moving through* the museum. Some, who would ordinarily move systematically through each room, instead breezed through, looking for objects that would catch their eye as a potential gift. Others abandoned their habit of looking quickly for things that they themselves would like in favour of a more methodical approach that would reduce the risk of missing out on something their friend might like. Thirty-three respondents spoke of the app as a type of 'guide' that gave focus to their visit. Some of the more negative responses we received described the app as nothing more than 'another tour' or as a distracting task (7 participants), but these were far outweighed by those who spoke positively about the experience. In fact, several spoke about the positive effects of having constraints put on their visit, and 23 users of it spoke in terms of the 'freedom' the app gave them.

Several visitor responses related to their sense of *connection* to the museum objects and to the person they were making their gift for (or receiving from). Just over half of those who responded to the question of whether they felt any connection to the objects in their gifts stated that they did. The few gift receivers we spoke to seemed to feel especially strong connections to their gifted objects, even when they themselves would not have found the object to be compelling on its own. Blast Theory's clearest achievement of a design objective, though, might be the connection that many visitors felt to their gift receiver or giver and the way that this connection was made possible by their interaction with the museum objects. Just over two-thirds of respondents indicated that having the app ask them to think about their receiver or giver as they looked through the museum created a sense of personal meaning for them. In fact, 20 respondents volunteered the terms 'emotional' or 'intimacy' in their descriptions, and several spoke movingly about shared memories. Most strikingly, when we asked people whether they felt a connection with the museum objects, many responded with descriptions of how connected they felt to their receivers or givers, not even realizing that they weren't answering the question they had been asked – because to them, the connection to the object was so directly tied to their connection with the other human being that there was no distinction to be seen between the two. The narrator's and/or giver's voice was named by 13 respondents as important to building the sense of connection.

One interview with a visitor, 'Mary', shows how her connection to her receivers affected their experience of the museum and of creating a gift. Mary describes how coming to the museum with her three children regularly throughout their childhoods has already shaped her relationship to the museum: 'When I come here on my own, I always want to talk to them about that time together.' Her gift contained three objects, one for each child. She described that by using the app, 'there was a much more personal connection to the objects'. The springboard for this connection to the museum was, in fact, her thoughts of her children. As she says, 'I could actually voice the emotional attachment I had to the museum and to the objects in the museum, and also to reflect on the changes that had occurred to the objects and their display, the curation'. She chose to photograph the balcony where they used to have tea together, a piece of fabric that her daughter used to love to touch, and a gruesome scene depicted as a ceramic that her young son used to gaze at 'with this horrified fascination, you know, and giggle at it almost, sort of: Did that really happen? Is that really here? And it's placed out of reach, just placed up here. And I feel that, again.' The fabric was not only a memory trigger but a marker of uniqueness and continuity within her family, because her daughter brought her own daughter to the museum with Mary, and without any prompting, Mary's granddaughter went straight to the same piece of fabric and touched it just as Mary's daughter had done.

When placed in the context of gift-giving, Mary's memories of her children brought her to vivid contact with those objects. And, in turn, she hoped that the app would transform the objects in the museum into the means to share something of these memories and positive emotions:

> I think, I hope they'll be touched by it. Yes, I just hope they would be. It's so nice to be able to record the sounds of the museum too, that ambient noise that you don't get in any other place, that sort of echoey voices and doors and that feeling that you're both alone and in a space with other people.

We acknowledge that some visitors found the idea of creating a museum gift to be too confusing or distracting, and either abandoned or would not choose to repeat the experience. This happened over all demographics and was independent of the level of familiarity with smartphone technology. However, many who were initially bewildered by the concept came to appreciate it by the time they had finished.

Most created a gift but very few received one during this trial. Some visitors arriving in pairs or small groups gave to each other, but the majority of testers chose to send a gift to somebody who could not receive their gift at the museum within the three days of the deployment. Therefore, we know far more about giving than receiving gifts using this app.

Analysis of the *Gift* App

As discussed in Chapter 2, the *Gift* app exemplifies a radical new way of thinking about personalization, which we have referred to as 'bottom-up personalization' or 'interpersonalization'.[4] So far, museum staff have generally borne the burden of 'modelling' their visitors so as to meet their needs and encourage visitor growth and/or loyalty. Modelling has been done by trying to anticipate their interests based on observations, questionnaires, preferences, familiarity, movement through the museum space, interaction with avatars, and a multitude of other techniques surveyed by Liliana Ardissono and colleagues, plus those that have emerged since then.[5]

The *Gift* app reveals how hybrid gifting helps to ease some of the burden of personalizing the museum visit by letting each visitor personalize it for themselves each time they come. More precisely, the *Gift* app gives a way for visitors to pay attention to the unique experience that they can make out of their visit by focusing on another person as the receiver. In other words, the *Gift* app exemplifies the interpersonalization discussed in Chapter 2.

It may seem as though Blast Theory's design would make the museum a less important part of the experience by directing the visitor's attention to their friend. But this design choice ultimately serves the needs of the museum by enabling new types of connection to both receiver (or giver) and museum. The giver or receiver does indeed act like a lens to create new ways of seeing and experiencing the museum, imbued with the positivity of intent lying behind the activity of thinking up a good gift for a loved one. The museum does not have to *do* the personalizing so much as *embrace* each visitor's unique personalization.

The rich descriptions that so many visitors gave us helped us to understand how they made sense of the *Gift* app and how they felt while using it. Chapter 9 goes into more detail about the types of automated data that can be generated, some of which opens up new opportunities for museum staff. In combination, we can learn more about the app as it is actually used. For example, less technically confident people are unlikely ever to try this app, as are adults responsible for multiple small children, old friends headed straight for the café – the list goes on.

Of those who do try the app, the value proposition or the tone are not to everyone's liking (as seen in the sharp drop-off during the introduction, shown in Chapter 9). But once visitors have invested in the project, we see a steady engagement. We can also infer that a significant percentage know they are giving to someone they do not expect will come to the museum

4 Ryding and others, 'Interpersonalizing Intimate Museum Experiences'; Spence and others, 'Seeing with New Eyes: Designing for In-the-Wild Museum Gifting'.

5 Ardissono, Kuflik, and Petrelli, 'Personalization in Cultural Heritage'.

in person, because they do not bother with leaving clues to the objects' physical locations.

Blast Theory were always keenly aware that these personal gifts were, essentially, digital media files that could be widely distributed without much effort. In fact, we anticipated that some visitors would want to share a gift they received by sending it to others privately or even posting it for all to see on social media sites. Most of our interviewees responded that they would not do that because the gift felt too personal to share. A few saw their gift as a future topic of reminiscence between themselves and their giver or receiver, and a few others said they might share with close friends or family members. This sentiment was borne out in the types of data explored in Chapter 9, which revealed no mass receiving at all, but a fair amount of gifting between pairs or small groups of friends. From the intensity with which so many spoke and the mass of automated data that supports their statements, we conclude that bottom-up personalization (or interpersonalization) may tap into the emotional motivations that underlie each person's visit in ways that more fully immersive technology-enabled museum experiences such as Mat Collishaw's *Thresholds* (see Chapter 9) do not seem to.

Analysing Blast Theory's design aims makes us look closely at the difference between sharing and gifting in a digital context. Their focus on sending personal content over commonly used digital communication platforms uncovers a very different approach to gifting than most digital exchange practices afford. A gift sent via the app *is* most often experienced as a gift, even though it consists of the types of photos and audio files that get transferred and shared every day.

The *Gift* app demonstrates that there is a clear difference in how sharing and gifting are experienced and in how people want digital experiences intended as personal gifts to be controlled. In design practices, though, gifting and sharing are very often muddled. This becomes particularly true in digital contexts, where the giver loses nothing in the exchange, often not even money. Not only have designers sometimes failed to create designs that are accepted as meaningful gifting, but it has been difficult to even express what the problem is.

The concept of inalienability emerged as a candidate.[6] Annette B. Weiner argues that the value of a gift partially resides in how the receiver perceives it as connected to the giver through the act of giving.[7] When applied to gift

6 Spence, 'Inalienability'.

7 Weiner, *Inalienable Possessions*.

transactions in a market-based economy, Weiner's 'inalienability' refers to the fact that thoughts of the giver are inextricably linked to thoughts or use of the gift, at least while memories of the giver and gift experience persist. A gift's inalienability lies in its power to evoke memories and emotions, or otherwise exert symbolic value. While the bulk of Weiner's research centres on cultural artefacts and the groups or individuals that hold them, she allows that the basic concept also applies to more commercial gift-giving practices.

The concept of inalienability helps us understand both the difference between sharing and gifting, and why it is important to differentiate the two. Thus, a digital or hybrid transaction in which the giver loses nothing and spends no money – as in online sharing – can be understood as a gift when what is 'shared' is mentally and emotionally inalienable from its giver. Moreover, inalienability is a matter of degree: A gift's inalienability can be as weak as a vague memory of having received it from some cousin or other at an unimportant holiday gift exchange event, or as strong as a deceased loved one's antique wedding ring. Turning museum objects into gifts becomes a case of the *Gift* app helping the visitor to tune into a meaningful relationship and digital media as the building blocks for attaching some inalienability to that object. Personalizing museum visits through a pleasurable, novel experience rooted in existing, meaningful relationships can be critical for cultural heritage institutions, who rely on visitor engagement to prove their relevance in an increasingly demanding climate of tight budgets and short-term metrics.

Conclusions

The *Gift* app is nothing more or less than what each individual user makes of it. There is no way any of us can even imagine funnelling each visitor into a single way of understanding and experiencing this app. Each visitor gets to make sense of it their own way. For some, that leads to long, intense periods of walking through the museum and studying every artefact, to give a well-explained, well-reasoned gift. For some, it led to a quick dash through a couple of rooms to find an object in their friend's favourite colour, record an intentionally silly message, and then continue a leisurely exploration of the museum without the app. Some people found the narration warm or soothing and perfectly paced. Others found it far too slow, even menacing. Some loved the chance to put themselves in another person's shoes (or eyes), while others saw it as the digital equivalent of a gift shop postcard.

These are all equally valid, equally valuable experiences, and no amount of design could force all these people into the 'same' experience. With its playful, personal approach, the *Gift* app lets people bring their own expectations, personalities, moods, and relationships into their museum visit – and that gives everyone the chance to create something new and malleable, even in cultural heritage institutions dedicated to preserving the ancient and venerable.

To date the *Gift* app has been used for another long-term, fully in-the-wild installation at the Brighton Museum (2019–2020); at the Munch Museum in Oslo, Norway (2019–2020); and the Museum of Applied Art in Belgrade, Serbia (2020). Talks are ongoing at the time of writing with Staatsgalerie, Stuttgart, Germany; the Design Museum, London, UK; the International Slavery Museum, Liverpool, UK; the Hong Kong Heritage Museum, Hong Kong, China; and Horsens Museum, Horsens, Denmark. And because a basic version is available for free it could already be in use at a museum near you, too.[8]

Bibliography

Liliana Ardissono, Tsvi Kuflik, and Daniela Petrelli, 'Personalization in Cultural Heritage: The Road Travelled and the One Ahead', *User Modeling and User-Adapted Interaction*, 22.1–2 (2012), 73–99 <https://doi.org/10.1007/s11257-011-9104-x>

David Cheal, *The Gift Economy* (New York ; Abingdon, Oxon: Routledge, 2015)

Bronislaw Malinowski, *Argonauts of the Western Pacific: An Account of Native Enterprise and Adventure in the Archipelagoes of Melanesian New Guinea* (London: Routledge (reprint 1978), 1922)

Marcel Mauss, *The Gift: Forms and Functions of Exchange in Archaic Societies* (London: W. W. Norton & Company (reprint 2000), 1925)

Julie A. Ruth, Cele C. Otnes, and Frederic F. Brunel, 'Gift Receipt and the Reformulation of Interpersonal Relationships', *Journal of Consumer Research*, 25.4 (1999), 385–402 <https://doi.org/10.1086/209546>

Karin Ryding, Jocelyn Spence, Anders Sundnes Løvlie, and Steve Benford, 'Interpersonalizing Intimate Museum Experiences', *International Journal of Human–Computer Interaction* (2021), 1–22 <https://doi.org/10.1080/10447318.2020.1870829>

Barry Schwartz, 'The Social Psychology of the Gift', *American Journal of Sociology*, 73.1 (1967), 1–11 <https://www.jstor.org/stable/2776124> [accessed 16 June 2021]

8 See https://gifting.digital or https://blasttheory.co.uk (Accessed December 2020)

Jocelyn Spence, 'Inalienability: Understanding Digital Gifts', in *Proceedings of the 2019 CHI Conference on Human Factors in Computing Systems*, CHI '19 (New York, NY, USA: ACM, 2019), p. 657:1-657:12 <https://doi.org/10.1145/3290605.3300887>

Jocelyn Spence, Benjamin Bedwell, Michelle Coleman, Steve Benford, Boriana N. Koleva, Matt Adams, and others, 'Seeing with New Eyes: Designing for In-the-Wild Museum Gifting', in *Proceedings of ACM CHI Conference on Human Factors in Computing Systems (CHI 2019)* (presented at the CHI 2019, Glasgow, Scotland, UK: ACM, 2019) <https://doi.org/10.1145/3290605.3300235>

Annette B. Weiner, *Inalienable Possessions: The Paradox of Keeping-While Giving* (Oakland, CA: University of California Press, 1992) <https://www.jstor.org/stable/10.1525/j.ctt1ppcbw> [accessed 23 June 2021]

About the Author

Jocelyn Spence is an interaction design researcher with an interest in performative experience design, performance, gifting, and cultural heritage; she is a Research Fellow in the Mixed Reality Lab at the University of Nottingham.

4. *Never let me go* – Social and Introspective Play

Karin Ryding

Abstract
This chapter presents a case study of a hybrid experience designed for art museums called *Never let me go*. It allowed art museum visitors to playfully guide a companion – using two interconnected mobile web apps and headphones. The purpose was to find a way to foster both social play as well as moments of introspection. The chapter gives insight into the design process and provides brief guidelines on how to design for play in general. The results from a trial that took place at the National Gallery of Denmark in Spring 2019 are described and discussed, particularly in relation to empathy and serendipity.

Keywords: Play; introspection; Interpersonal relation; Meaning-making; Empathy; Serendipity

Museums have a dilemma. On the one hand, museums wish to offer their visitors experiences that are profound, meaningful and contemplative – what has been called transformative experiences.[1] On the other hand, as has been pointed out in Chapter 2, museum visits are rarely solitary experiences. Rather, they are often highly social, as visitors come in pairs or in groups, and use their visit both to spend time with the people they come with as well as sharing their experience through social media. This leads to a dilemma: How can museum experiences simultaneously live up to the transformative ideal, while also taking into account the social dimension? This chapter explores this dilemma through the design of a playful museum experience called *Never let me go*, which aims to facilitate

1 Soren, 'Museum Experiences that Change Visitors'.

Waern, A. and A. Sundnes Løvlie (eds.), *Hybrid Museum Experiences: Theory and Design*. Taylor & Francis Group, 2022
DOI 10.5117/9789463726443_CH04

an experience that is at one and the same time social, while also directing the participants towards an introspective encounter with the museum collection.

Never let me go is a hybrid experience designed specifically for art museums. The aim of the design was to create a playful experience for adults that would work as a form of interpersonalization (see Chapter 2). Therefore, it was designed for two persons to play together in a way that would reshape the museum experience significantly by making it more intimate, personal and emotional (as well as more playful of course).

The name, *Never let me go*, gives a little hint about the experience: It can be described as an open-ended game in which one player relinquishes control over their actions and overall experience to someone they trust. This way, it invites the players to engage in creative and immersive play as they explore the museum together. The design fosters light-hearted social play as well as deeper moments of introspection, and by doing so it highlights some of the potentials of interpersonal hybrid museum experiences. It also illustrates an approach that to a large degree favours visitor control over the experience, rather than curator control. At the same time, the museum environment and exhibitions remain very much central to the experience.

The chapter gives an insight into the design process of *Never let me go* as well as a detailed description of the final version of the prototype. Finally, the results from a trial that took place at the National Gallery of Denmark in the spring of 2019 are described and discussed.

Becoming Mouse: The Experience

You are visiting an art museum with your partner. As you enter the exhibition, you take up your phone and activate the *Never let me go* web app, and enter a code to connect with your partner. You put your headphones on, and you place your phone in your pocket, because this time you will be the Avatar. You feel excited and a bit nervous. What will she tell you to do? Where will she lead you? What will you experience? Your partner looks at you to check if you are ready. You nod. She smiles and then presses a button on her phone. Suddenly you can hear a calm female voice saying:

> Welcome to this avatar experience. You will soon hear instructions chosen by your partner. Follow these instructions to your own ability and desire. Make it as dramatic or as subtle as you wish. Remember to stay safe and

stop whenever you want. When in doubt of what to do, relax and enjoy the art. Now start by doing just that. Enjoy!

After a moment you start moving, a bit hesitatingly at first. Then you hear the voice saying 'Explore' and you begin to walk more purposefully through the room. It is a big room full of sculptures in different sizes. You stop and look at a sculpture of two rather disturbing looking animals. The voice asks you 'What does it remind you of?', and you start thinking of a trip to the zoo when you were a young child. It always made you sad to see the animals locked up as they were.

You look at your partner. Did you tell her this memory before? You wonder now what she is thinking. The voice says, 'Can you sense the anger in this?' and you stop for a while to take this in. Yes, there is anger, definitely, but also so much sadness. Your emotions make you pause for a moment, then you move on to the next sculpture. This time it is a tiny sculpture of a mouse looking up at you. The voice says: 'Become small'. You wonder for a moment what to do, but then you crouch down next to the sculpture. It looks different like this. You can see all the details now: The fur, the tiny ears, the black, beady eyes.

The voice says, 'Imagine this is looking back at you'. You feel like you are watching the mouse and the mouse is watching you. As you sit there on the floor watching the mouse intensively, you feel a sort of connection with it. The voice now says, 'Become part of this'. To your own surprise, you do not hesitate long before trying to become a mouse yourself. As you attempt to shape your body into an odd, mouse-like posture, you suddenly become aware that your partner has sneaked up on you and is now crouching down right beside you. You look at her with big eyes, like the mouse you imagine, and you make a rather shrilling 'piiii' sound. It comes out a bit louder than intended, and you both start to laugh. You laugh until your stomach really hurts and you cannot laugh any longer. When you have finally calmed yourselves down a little, your partner gives you a quick kiss and then presses a button on her phone. The voice says, 'Follow me', and you stand up to get ready for the next part of the adventure.

Designing for Playfulness

Play is a very specific way to be in the world. It gives us room to explore limitations as well as possibilities in movement, ideas, affects or whatever we choose to play with. Play relies on structures such as rules and frames, but just as much on players' spontaneous and creative expressions within those structures.[2] As Sicart has put it, play is not necessarily about having fun. It is about 'opening us to the immense variation of pleasure in this world'.[3] It is in essence an appropriative behaviour, which includes trivialization as well as making things deadly serious. It facilitates meaningful social connections.[4] Moreover, play can empower us by expanding our feelings about what we can be and do in the world.[5]

Never let me go was designed as a social tool to orchestrate playful experiences in art museums. This setting was chosen precisely because playful behaviour is not what we typically associate with the traditional art museum visit. In these environments, most people would walk slowly and quietly, stopping now and then to look at the artworks. Carol Duncan compares this behaviour to following a script.[6] According to her, the museum's sequenced spaces and arrangements of objects, its lighting and the architectural details provide both the script and a stage set. Visiting an art museum can in this way be seen as taking part in a form of secular ritual. The situation resembles in some respects certain medieval cathedrals where pilgrims followed a structured narrative route through the interior, stopping at prescribed points for prayer or contemplation.

Play on the other hand rarely follows a script but has its own rules. When we play, we accept the rules of the game and by doing so we allow ourselves to act in ways we perhaps wouldn't do outside of play. This protecting boundary of play is what play scholars sometimes call 'the magic circle of play'.[7] It basically consists of a social contract between the players, which is often decided on beforehand but can also be negotiated during play. Because players often experience that the contract protects them to some degree from awkwardness or embarrassment, play has the potential to work as an alibi for players to try out new ways of being for a short period of time.

What does it mean to design for play? Play is something that happens spontaneously, and it is hard to predict. Sometimes we feel like playing and sometimes not. The playful attitude can't be pushed onto people, it needs to come naturally. There are ways to achieve this through design, but there is

2 Caillois, *Man, Play and Games*; De Koven, *The Well-Played Game: A Player's Philosophy*.

3 Sicart, *Beyond Choices: The Design of Ethical Gameplay*, p. 3.

4 Isbister, *How Games Move Us: Emotion by Design*.

5 Henricks, *Play and the Human Condition*.

6 Duncan, 'The Art Museum as Ritual'.

7 Stenros, 'In Defence of a Magic Circle: The Social, Mental and Cultural Boundaries of Play'.

not one solution that will fit all. Bill Gaver and his colleagues suggest three general guidelines when designing for ludic activities.[8]

1. Promote curiosity, exploration and reflection.

In order for play to happen there needs to be an element of uncertainty and discovery. People need to be allowed to explore, tamper with things and figure them out for themselves.

2. De-emphasize the pursuit of external goal.

Interestingly, if a system can easily be used to achieve practical tasks, this will actually distract from the possibilities it offers for more playful engagement. This means that if you design an app which is primarily a museum guide, it will be mainly used for this purpose. Any playful elements that might be included in the app will perhaps not even be noticed by its users.

3. Maintain openness and ambiguity

If there is too much structure or predefined meaning to an activity or a design, it will inevitably suppress play. If we want people to engage playfully, there needs to be a certain amount of ambiguity and open-endedness to the design in terms of how people may interpret it and give it meaning.

Designing *Never let me go*

Throughout the design process for *Never let me go* the following question was asked: What kind of play would not only be possible, but would also feel meaningful, in an art museum setting? There is of course not just one answer to this question. However, as the design process progressed two main criteria for the experience emerged. The first was that it should be social and support close relationships. The second was that it should be a playful experience that included moments of introspection. The two criteria are based on research that shows that the most satisfying experiences during a museum visit are not just object-related (such as seeing the 'real thing') and instructive but also *introspective* (focused on imagining, reflecting,

8 Gaver and others, 'The Drift Table'.

reminiscing and connecting) and *social* (interactions with friends and family).[9]

As a result, *Never let me go* was designed as a two-player game which lets players take the roles of the Avatar and the Controller. The Controller is given the tools to spontaneously orchestrate an experience for the Avatar, taking place in real time in the museum. The system is designed as two connected web apps where only the Controller app has an interface. The interface consists of a menu with different prompts, questions or instructions that can be sent to the Avatar, who will receive them as pre-recorded voice messages. The Controller shares audio with the Avatar in order to keep track of how the experience is playing out. The shared audio is also used to emphasize intimacy and to create a shared space where the two participants can feel safe together.

The content in *Never let me go* is designed to be building blocks for emergent, playful experiences. It offers a structure with a clear beginning and an ending to set a frame for both players to act within. The intention is for players to use the content in an open-ended way. Therefore, a variety of content is implemented which can be interpreted differently depending on the situation. In this way, openness and ambiguity are used to give room for curiosity and exploration. This also allows for users to express themselves and decide upon which tone to set on the experience. The idea is that this enables both fun as well as serious encounters with the art and each other.

In the Controller app, there are six different categories to choose from in the menu. The first is called 'Basic commands' and consists of direct prompts such as 'Explore', 'Follow', and 'Wait'. The second is called 'Body' and consists of instructions relating to the body of the Avatar, such as 'Close your eyes', 'Breathe deeply' or 'Mimic this with your body'.

The third category consists of personal questions that can be used in relation to the art, for example 'What part of your life is connected to this?' and 'Who would you give this to?' The fourth category is called 'Becomings' and consists of prompts that are open for interpretation. Examples are 'Become light', 'Become sharp' and 'Become part of this'. The fifth category is 'Feelings' which consists of questions again to be related to the artworks, but this time in order to direct the Avatar's attention to the emotional content of an art piece. Examples are 'Can you feel the longing in this?' or 'Can you sense the pain in this?'.

9 Pekarik, Doering, and Karns, 'Exploring Satisfying Experiences in Museums'.

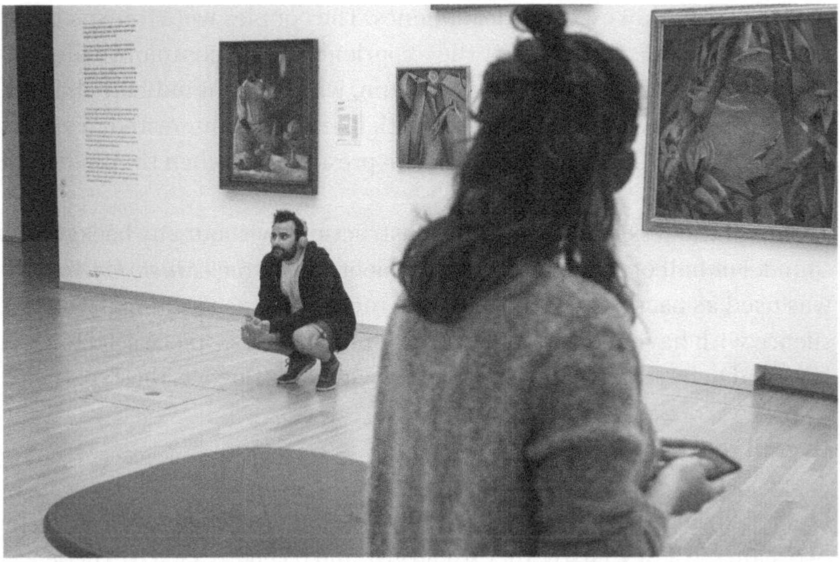

Figure 4.1: From the trial at the National Gallery of Denmark. Photographer Johan Peter Jønsson.

Lastly, there is a category called 'Imaginings'. This consists of instructions with the intention to trigger the Avatar's imagination. The idea is also to induce a sense of urgency in order to intensify the Avatar's experience. Examples of this category are 'Imagine that everything here is about to fall apart' and 'Imagine that this is looking back at you'.

Apart from the categories, there are 'Begin' and 'End' options in the menu. These trigger longer instructions, directed at both the Avatar and the Controller.

The Visitor Perspective

The main trial of *Never let me go* took place at the National Gallery of Denmark around Easter of 2019 (Figure 4.1). Twenty (20) persons of fourteen different nationalities signed up to take part. They were recruited through social media and through a mailing list for people interested in cultural events in the Copenhagen area.

Each test was separated into four different sessions, each approximately ten minutes long. After a session ended, the participants would swap roles. Thus, everyone would try out both the Avatar role and the Controller role twice, in order to allow us to observe how they progressed in their use of the roles. Before starting, all participants were equipped with a mobile device and a set of over-ear headphones. The couples were free to choose where in the museum to start the experience. Most commonly this was in the modern art section of the museum, which might indicate that this was the section which most of the participants were interested in seeing. The first Controllers were instructed to press 'Begin' when they felt ready to start.

Half of the sessions had only the instructions without any background sound. For half of the sessions, the ambient soundtrack *Music for Airports* was used as background music.[10] The intention was to compare having silence with having music as a background. The music piece selected was intended to be relaxing but not interfere with, or influence, the experience.

Overall results

From the trial, it became clear that the participants used *Never let me go* for two main purposes. Firstly, they would give and receive personal experiences in relation to the art and the museum space, and secondly, they would explore their personal relationship through playing, teasing and pushing social boundaries.

More than once, the experience was described as *being in a bubble together.* Playing with the app became a new way of communicating during their visit, different from how the participants would normally interact with each other. When describing the experience, the players would refer to a prompt being sent or received by saying 'I said', 'she said' or 'he said'. The fact that they had used someone else's voice to communicate didn't seem to matter. One player compared it to having a secret language – a language that enabled them to say things silently. This became evident through the fact that players most often chose not to speak to each other directly at all during play. Instead they used body language to communicate the things that they couldn't say using the system. In a few cases, though, the players would simply remove their headphones to talk, for example to answer a question that had been sent and received. The silence, or the lack of ordinary conversation, seemed to be particularly enjoyable to some of them. As one female player says,

> Being able to communicate without having to come up with the things to say. Not having to talk. It was so relaxing. I really love that.

10 Eno, *Ambient 1 – Music for Airports.*

Being an Avatar

Being an Avatar was both relaxing and had a certain tension to it. Players reported experiencing a sense of freedom, but at the same time they would be anticipating the actions of the Controller: 'I remember enjoying being on my own, doing my stuff. But also, this tension and anticipation about waiting for commands and how that was actually very interesting.' The passive, receiving quality of the Avatar role could also be frustrating to some people.

All players felt a strong obligation to follow the prompts they received, and they would feel guilty if they were not able to respond appropriately. Nevertheless, for their own amusement Avatars would sometimes take the liberty to knowingly misinterpret a command. They would twist the meaning somewhat and do what was suggested but not in the expected context. A player gives this example:

> You came next to me and said: 'Come closer'. I knew, I was sure that you meant go closer to the painting, but I thought I'm not going to go closer to the painting. I'm going to go closer to her and make her uncomfortable. That was fun!

This type of behaviour was part of how the players would often make jokes, play and tease one another. Being the Avatar could also be a bit challenging in this regard. The physical prompts were often used by Controllers to push the Avatars to the limit of what they were willing to go along with. This would not surprisingly sometimes lead to resistance. As a player explains, 'For example, she would tell me to stretch or mimic. And I did it a few times but then I was feeling very awkward. So, I didn't.'

What most participants enjoyed was how the different prompts, particularly the questions, would trigger personal memories and fantasies. A player here describes a situation where he was standing in front of a painting depicting a view over the ocean:

> So, when I got the question 'Where are you?'. I would have expected the answer to be like, I'm right here! But that wasn't my experience. I went to where does this painting actually take me. And it took me to a summer holiday trip where I remember I was standing at the beach and looking at the waves.

Even abstract paintings could trigger this type of experience, as a player describes:

> I really enjoyed those colours, the blue and the green. And when the question was 'Can you see yourself in this?' I could see maybe the difficult moments in the spikes. I just let my imagination go into that abstract painting.

This led to deeply personal moments for the Avatars especially, but sometimes also for the Controllers who would be inspired to engage emotionally.

Being a Controller

Being a Controller was, perhaps not surprisingly, a very different experience from being an Avatar. This role was much less relaxing. As one player explains,

> I felt a lot of responsibility. Both for the person's safety, but also for the person's enjoyment of the experience, and also the artworks.

Taking on this role was interpreted as taking on the challenge to build meaningful and cohesive experiences for another person. In order to do so, most Controllers tried to be aware of the actions and whereabouts of the Avatar, at the same time as thinking of the artworks and what interpretations they offered. As the same player puts it,

> As a controller I'm trying to interpret what the art is or gives me, in order to give that to the person who is looking at it.

In general, there were three different approaches taken by the Controllers. Often, they would observe the Avatar closely, sending a prompt only when they thought they could add to or enhance the other person's experience. This seems to have worked well for the Avatars, but the Controllers themselves sometimes felt frustrated when they didn't get any reply or indication of how their prompts had been received. At other times, Controllers sought to dominate the experience by pulling their Avatars away from what they were doing or pushing them to act in ways they wouldn't choose themselves. A few times, Controllers would let go of the control completely and instead send prompts at random, leaving it up to the Avatar to make sense of the situation.

An Exercise in Empathy

Empathy is a social process. It happens as we become aware and understand someone else's emotions and/or behaviour and it often leads to us experiencing the same emotions as the other person. Taking someone else's perspective, or 'seeing things through their eyes', is a well-known precursor to feeling empathy.[11]

During the trial of *Never let me go* players were given Galvanic Skin Response (GSR) sensors to wear on their left hands. In this way, it was possible to track the player's emotional arousal during play. What the GSR sensor does is to measure the activity of the sweat glands in the hands – a bit similar to how a lie detector works.[12] It is not possible to measure which emotions players are experiencing, only how intense they are. Therefore, it is vital to combine the GSR data with data from interviews with the participants in order to get an understanding of what they were actually feeling at the time.

The purpose of using GSR tracking was not to verify the players' interpretations of the events, but rather to use it in an exploratory fashion; to look for patterns in order to gain a richer understanding of this type of interpersonal hybrid experience. It is important to point out that there are several challenges with using GSR tracking outside of a laboratory environment. Disturbing elements, such as sudden sounds or movements that may happen inside a museum, cannot be excluded during the trial. The GSR data will therefore include emotional reactions to all kinds of stimuli outside of the designed experience. Nevertheless, what could be done in this particular case was to compare the GSR data from the Avatars and the Controllers to look for patterns. For the researchers, this gave some interesting insights into the interpersonal aspects of the experience.

Several of the participants described the Controller experience as being so focused on the Avatar that they would feel what they assumed the Avatar was feeling. As a player puts it:

> It is an interesting and engaging experience to be the controller. It forces you to put your attention on the other person, and try to be doing an empathy exercise. Putting yourself in the shoes of the other.

11 Batson and others, 'An Additional Antecedent of Empathic Concern: Valuing the Welfare of the Person in Need'.

12 Sharma, Kacker, and Sharma, 'A Brief Introduction and Review on Galvanic Skin Response'.

The occurrence of this 'empathy exercise' could be discerned in the results from the GSR tracking. From the sessions where valid data could be retrieved from both roles, the patterns of emotional arousal showed clear similarities. What these findings suggest are that when people spontaneously create or orchestrate a personal experience for someone they care for, they will often get as emotionally involved in the experience as the person receiving it.

Intimacy as a Resource

The set up with the two roles, in which one was responsible for the other, established a level of intimacy between the players of *Never let me go*. This did not only lead to feelings of empathy between them. Interestingly, it also led to an increased level of intimacy with the artwork. The receptive quality of the Avatar role helped players to become more attentive and to experience the art in new ways. A player explained it this way:

> I think it was a chance to connect with the art and not just be an observer, but to be part of the paintings but also the whole room. It helped me enjoy it and understand it more. And think about it more. It wasn't just my eyes watching. It was my whole mind observing.

In this sense, the hybridity of the experiences added a structure inside which participants could explore a variety of different ways of being in the museum. This ranged from light-hearted teasing and playing, to embodied explorations of the art and the architecture as well as deep introspection. The level of intimacy in combination with the possibility to act spontaneously even led to a few occasions of serendipity. Serendipity can be described as an experience where conditions seem to fall perfectly into place in a surprising, almost magical, way.[13] This rarely happens in our everyday lives, but when it does it leads to a change in a person's state of awareness, coupled with a positive emotional reaction.

This is an example that was reported during the trial. An Avatar is standing in front of a large window to enjoy a view over Copenhagen. Exactly at the same time as he is looking directly at the street where his girlfriend is about to open a gallery of her own, she, as the Controller, sends him the prompt 'Imagine that this is the beginning of something new'. It is of course not possible for her to read his mind, but by knowing him intimately she

13 Makri and Blandford, 'What is Serendipity?'

is able to act on the spur of the moment when the opportunity presents itself. This resulted in the Avatar feeling like the experience was tailored uniquely for him, and as a consequence it made him even more emotionally invested in it.

Both empathy and serendipity can lead to strong emotional experiences. The occurrence of them during the trial of *Never let me go* points to the potential of designing tools for visitors to playfully and spontaneously orchestrate intimate hybrid museum experiences for their loved ones.

Analysis of Never Let Me Go

As was discussed in Chapter Two, one the main potentials of bringing play into the museum is the ability it gives participants to resignify and reinvent museum objects and spaces together. This can be done through experiences in which players enter a different state of mind; one that lets them focus more on their personal interpretations of the museum. This was one of the things that made the *Gift* app discussed in Chapter Three successful, as it let participants see the museum through the eyes of a close one. In *Never let me go*, a similar technique was used, as players would in a sense 'give' an experience to one another. However, due to the more playful and open-ended components in the design of *Never let me go*, a broader spectrum of player behaviour can be found in the results from study. As discussed above, several players describe having profound, and potentially transformative, personal and emotional experiences. At the same time, participants would often use *Never let me go* as a chance to play and make internal jokes, using the artworks as props in their personal narratives. As one player puts it, 'if I can't connect emotionally with the art, then I can have fun with it'.

There may be ethical implications in supporting this kind of playful behaviour in an art museum. Taking over spaces and pushing social boundaries are part of play and therefore, as Sicart argues, it exists in a tension between creation and destruction.[14] Allowing for play means losing a certain amount of control over visitors. *Never let me go* gave participants an alibi to do things they wouldn't normally do when visiting an art museum. They laughed, put themselves in awkward bodily postures, followed strangers, hid from each other, and went looking for things they could touch. Many of the players reported feeling a sense of freedom; however, they also described

14 *Beyond Choices: The Design of Ethical Gameplay.*

how they were completely aware at all times of both the social and legal boundaries of the museum. They would push each other to do things, but always making sure not to disturb other visitors or to get into trouble with the guards. In this sense, it became clear that the players knew exactly where the lines were between acceptable and non-acceptable behaviour. One could therefore argue that, when it comes to the average adult art museum visitor, encouraging playful behaviour is not putting the museum at risk (in terms of inappropriate behaviour, vandalism, etc.). Instead, it enables visitors to find new, more embodied, perhaps unexpected, ways to encounter the art.

Conclusions

The motivation behind this study was to show that play is possible even in an environment that is not traditionally associated with this kind of behaviour, namely the art museum. Museums tend to create experiences for their visitors that follow linear narratives or offer interactions that are more or less predetermined. These experiences can indeed be enjoyable or inspiring, but what is suggested here is that new potentials can come out of giving more control over to the visitors themselves.

For museum professionals, the main take-away from this case study is that what people look for in a museum is not only to learn new things about the world outside of themselves, which can be found in art, history or science, but what they value just as much is to learn new things about themselves and the people they are closest to. Designing for playfulness in ways that suits the specific environment of a museum can be a way to fulfil these different needs. It can lead to visitors having experiences that are both fun and social as well as deeply serious.

The results from this study showed that when people were allowed to play in an art museum humour was definitely an important part of it, but serious moments of introspection, reflection and emotional engagement were just as present. What the players valued mostly was the distinctly personal approach to the art, especially in combination with the ability to have a special connection to their partners. Play is always situational, which means that the specificities of the environment and the players matters. In this case, it gave the participants the possibility to gain new perspectives of the artwork of the National Gallery of Denmark, but also to get to know their companions better.

The *Never let me go* web app has been released as a prototype with the intention to inspire further development. More information about *Never let me go* and how it can be tried at your museum can be found on the GIFT website.[15] If you are interested in more guidelines or inspiration on how to design for playfulness, we would encourage you to take a look at the Playful Experience (PLEX) framework.[16] It contains 22 categories of playful experiences which can be used as a starting point in the design process.[17]

Bibliography

Juha Arrasvuori, Marion Boberg, and Hannu Korhonen, 'Understanding Playfulness – An Overview of the Revised Playful Experience (PLEX) Framework', in *7th International Conference on Design and Emotion* (presented at the 7th International Conference on Design and Emotion, Chicago, IL: Design and Emotion Society, 2010) <https://doi.org/10.5281/zenodo.2596777>

CD Batson, JH Eklund, VL Chermok, JL Hoyt, and BG Ortiz, 'An Additional Antecedent of Empathic Concern: Valuing the Welfare of the Person in Need', *Journal of Personality and Social Psychology* (2007), 65–74

Roger Caillois, *Man, Play and Games* (Champaign, IL: University of Illinois Press, 2001)

Bernard De Koven, *The Well-Played Game: A Player's Philosophy* (MIT Press, 2013)

Carol Duncan, 'The Art Museum as Ritual', in *Civilizing Rituals* (London: Routledge, 1995) <https://doi.org/10.4324/9780203978719>

Brian Eno, *Ambient 1 – Music for Airports* (E.G. Records Ltd., 1978)

William Gaver, John Bowers, Andrew Boucher, Hans Gellerson, Sarah Pennington, Albrecht Schmidt, and others, 'The Drift Table: Designing for Ludic Engagement', in *CHI '04 Extended Abstracts on Human Factors in Computing Systems*, CHI EA '04 (New York, NY, USA: ACM, 2004), pp. 885–900 <https://doi.org/10.1145/985921.985947>

Thomas S. Henricks, *Play and the Human Condition* (Champaign, IL: University of Illinois Press, 2015)

Katherine Isbister, *How Games Move Us: Emotion by Design*, Playful Thinking (MIT Press, 2017)

Stephann Makri, and Ann Blandford, 'What is Serendipity? A Workshop Report', *Information Research*, 16.3 (2011) <http://www.informationr.net/ir/16-3/paper491.html> [accessed 25 June 2020]

15 http://gifting.digital/

16 Arrasvuori, Boberg, and Korhonen, 'Understanding Playfulness – An Overview of the Revised Playful Experience (PLEX) Framework'.

17 The PLEX framework is also available as design cards: funkydesignspaces.com/plex/ (Accessed December 2020).

Andrew Pekarik, Zahava D. Doering, and David A. Karns, 'Exploring Satisfying Experiences in Museums', *Curator: The Museum Journal*, 42.2 (1999), 152–173

Mahima Sharma, Sudhanshu Kacker, and Mohit Sharma, 'A Brief Introduction and Review on Galvanic Skin Response', 2.6 (2016), 13–17 Journal title is missing. <https://doi.org/10.21276/ijmrp.2016.2.6.003>

Miguel Sicart, *Beyond Choices: The Design of Ethical Gameplay* (Cambridge, MA: The MIT Press, 2013)

Barbara J. Soren, 'Museum Experiences that Change Visitors', *Museum Management and Curatorship*, 24.3 (2009), 233–251 <https://doi.org/10.1080/09647770903073060>

Jakko Stenros, 'In Defence of a Magic Circle: The Social, Mental and Cultural Boundaries of Play', in *2012 International DiGRA Nordic Conference* (presented at the DiGRA Nordic '12, Tampere, Finland: DiGRA, 2012), 1–19

About the Author

Karin Ryding is a PhD Fellow at the IT University of Copenhagen, Denmark. Her research interest includes critical play, game design, affective interactions, posthuman perspectives and cultural heritage. She holds an MA in Visual Culture Studies from Lund University and a BA in Game Design from Uppsala University. Outside of her research career, Karin has worked with designing board games, locative games, transmedia and other forms of experimental and creative play experiences as part of Ozma Games.

5. *Your Stories* – A Life Cycle Analysis

Paulina Rajkowska

Abstract

This chapter deals with another challenge for hybrid museum experiences: How can the museum ensure that a hybrid experience continues to work well once in place, throughout the time that the system is in place? Through a study of an augmented reality installation in the National Museum of Serbia, Belgrade, the chapter discusses the many practical and organisational challenges involved in creating and maintaining a hybrid experience over time.

Keywords: Augmented reality, Co-creation; Meaning-making, History

Your Stories is a hybrid museum experience that was created by the interactive design company NextGame in Belgrade, Serbia and supported technologically by Nottingham University. The project was run in collaboration with the National Museum of Serbia, an important educational and research institution that had recently reopened their main exhibition building after ten years of renovations and collection changes. The core idea behind *Your Stories* was to invite visitors to contribute their own stories to the museum. A public call online resulted in multiple donations of objects that were then 3D scanned and documented by our team. Over 6 months of collaboration time, the museum and NextGame used those items to develop a hybrid installation in which the digital versions of everyday objects of today were paired with historical objects in the exhibit. Bringing modern stories into the museum helped re-contextualize inanimate objects in the gallery, showing how much meaning there can be behind small items and how perhaps the old items we admire at the museum once had similar value to people contemporary with them.

This study is an account of a life cycle analysis of the *Your Stories* experience. By this, we mean that our study aimed to follow the design project

Waern, A. and A. Sundnes Løvlie (eds.), *Hybrid Museum Experiences: Theory and Design.* Taylor & Francis Group, 2022
DOI 10.5117/9789463726443_CH05

from its creation throughout the development, implementation and testing, but also studying what happened once the research and development team had left, and the experience took on a life of its own. Research projects (including most of the designs pursued within the GIFT project), tend to focus on a particular research question and explore it during a limited timeframe, and end when the research has been concluded, often through an evaluation done at an early trial of a technology solution in the intended context.[1]

Since *Your Stories* was adopted by the museum as a permanent installation in their space, it allows us to look at the process in its entirety, considering the local context and the uses and experiences that emerged. *Your Stories* is a case of a technology development process done in a research context, that was adopted by the museum and maintained beyond the project partners' direct involvement. In this case, the experience remains available to visitors on a regular basis at the time of writing this chapter. That is a rare positive result which we believe is also connected to aspects of the design process which we will discuss further below.

The National Museum of Serbia

In contrast to the case studies presented in Chapter 3 and 4, *Your Stories* was designed for a specific museum. The National Museum of Serbia was established in the mid nineteenth century and survived the complicated times of the Yugoslav Wars as well as the Milosevic era. The physical museum was then closed in 2003 for necessary renovations and remained closed for fifteen years. It used to be a popular gathering spot for locals and the lack of the museum in the cultural life of the Serbian capital was notable. That said, the National Museum functioned over the years as an important organization promoting local cultural heritage, expanding their exhibits and focusing on research as the restoration of the building went on. The official grand reopening took place on the 28th of June 2018, coinciding with the Serbian National Day celebrations. It greeted the visitors with a freshly renovated facade, new exhibits as well as a variety of digital content including hybrid installations as well as VR solutions. *Your Stories* was picked up as a project shortly before the opening; practical work took place starting July 2018 and the exhibit was launched a few months later.

1 Taylor, 'Leaving the Field'.

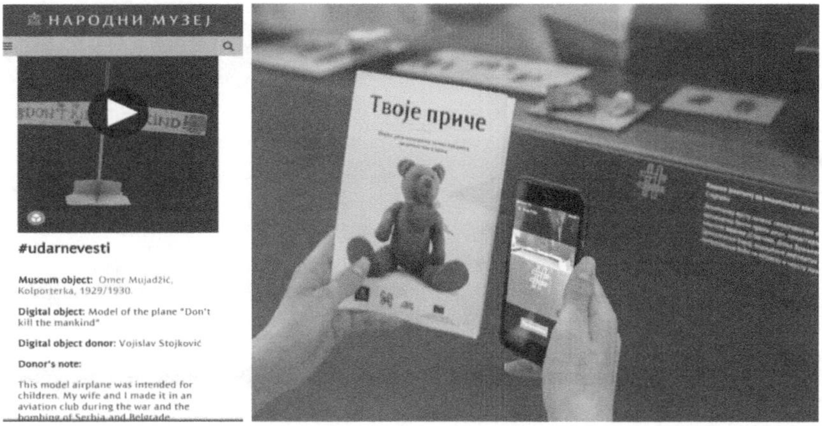

Figure 5.1. Left: Playable 3D model and description. Right: Visitor scanning a code and interacting with the exhibit according to instructions in the pamphlet. Images by NextGame.

The Design

Your Stories is an Augmented Reality (AR) experience using Artcodes[2] technology and 3D photogrammetry (the latter allows for the scanning of physical objects to produce 3D models of the same). The application allows the visitors to go through the museum space while hunting artcodes placed around the different exhibits. To begin the experience, the visitors receive a pamphlet which briefly describes how to access the application as well as containing a map of the codes in the entire museum. As the visitors approach the different exhibits they can see the characteristic symbol of an artcode, in this case shaped as a hashtag to represent the topical connection between online content and physical exhibit (Figure 5.1). Once scanned, the artcode shows the visitors a 3D model of an item that relates to the exhibit and which has been donated by locals, accompanied by a story that was provided by the donor. This pattern of interaction is identical for all of the objects.

The Design Process

The design of the app, as well as the selection of virtual and museum objects, was done in close collaboration between NextGame, the app designer, and the museum curators. The close collaboration was vital for the results

2 https://www.artcodes.co.uk/

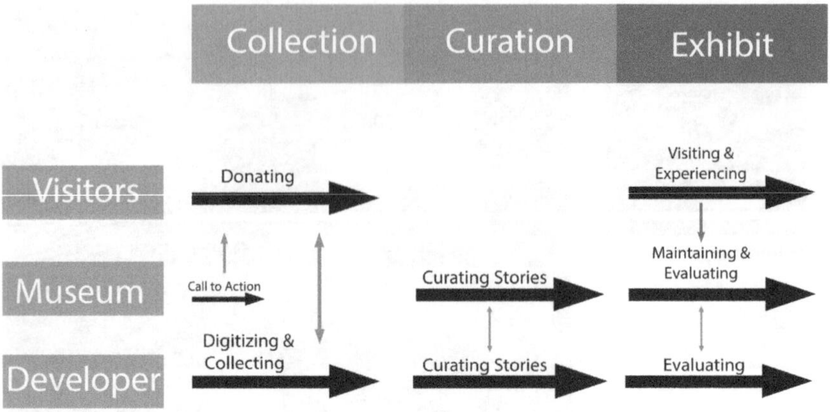

Figure 5.2. Work process of the exhibit *Your Stories*.

achieved. Constant communication, expectation management and working together contributed to the satisfaction level of the included partners.

The process can be roughly described in three stages as visualized in Figure 5.2. As shown in the figure, the different parties involved in the process collaborated in different intensities throughout the project phases. Here, we break down these different phases to give the reader some background on what took place during the design process.

Collection

The process began with a public invitation for visitors to bring their own important items to the museum, to get them 3D-scanned so that they could be added to the museum's online collection. The call was issued through the museum's official website in Serbian, to promote local engagement. The call encouraged museum visitors to bring their own items and stories with them, to entrust them to the museum collection. This process was technologically enabled through collaboration with University of Nottingham, and use of their 3D scanning technology. The scanning was done in multiple sessions over a period of 3 months starting in July 2018. Each visitor provided a story on what the object meant to them and to whom it had belonged.

Curation

Next, the developer and the museum reviewed the digitized items, together with the stories accompanying them, and started matching this new content with the existing physical artefacts at the museum. This became a lengthy process of curation and research, lasting over 6 months, done in collaboration by NextGame and the museum.

The time-consuming part was finding meaningful locations and associations for the donated digital content among the existing exhibits. As reported by both museum and the developer, the process required great care. It was difficult to find connections that were both relevant from a museum perspective, and would make sense to the general public. The museum wanted to make sure that the connection between the story and the item was clear and did not leave the visitors wondering how to interpret it. Matches were made, internally tested and remade based on feedback. It took many iterations as well as historical research. For example, one of the visitors had donated a baby shoe. This shoe was matched with an artefact from the stone age which was a children's toy from that time. It was one of the stronger connections that museum visitors praised during our evaluation. The experience was initially only available in Serbian, but an English version was developed shortly thereafter.

The reader should note that the original contributors themselves were not involved in this part of the process. That said, the raw content provided needed to be handled by the developers and the museum and rendered more legible to a general audience, and for this purpose the professional curatorial expertise was critical.

Once the items had been matched, an interface was designed using the Artcodes application, a piece of technology originally created by Nottingham University based on image recognition from the phone camera. This technology was chosen as Artcodes provides a reliable scanning technology that allows for graphical markers that are meaningful as well as visually appealing.

Exhibit

When the experience had been finalized, it was installed in the museum and launched at an inaugural event open to the general public. The visitors who had donated the objects were specially invited, this time to see what others had donated and to enjoy the stories created by associating their contributions with the physical exhibition. At the same time and after this launch, the National Museum and NextGame continued to work on maintaining the digital infrastructure as well as on developing marketing and instructional materials such as flyers and roll ups.

Visitor Perspective

Study

Both the Museum and the developer partner NextGame have performed evaluations of *Your Stories*. The Visitor perspective discussion below is primarily based on an evaluation carried out on behalf of NextGame, in which the chapter author took part. This testing took place in April 2019 and included 5 days of fieldwork at the museum, where we conducted in-depth interviews with visitors who agreed to test the application. The participants were recruited from regular museum visitors on the specific days, which resulted in a mix of international and local visitors. We used group interviews and individual interviews depending on the social context the visitor was in, to minimize the strain put on potential friend groups or families being at the museum together. At the time of writing this chapter, the museum's evaluation was still ongoing.

Findings

Your Stories would not exist without deep involvement of the museum visitors in the process of creation. Their willingness to volunteer their possessions and their stories were key to this idea taking off. The experience offered different levels of engagement to different audience members. On one hand, we have visitors that contributed to the creation of the exhibit, and on the other we have the visitors who experienced the stories of others during their stay at the museum.

Being able to add your personal possession to the museum collection was very appealing to visitors who volunteered their private time and effort. They did not receive any compensation for the work, but they were invited to the official opening of the exhibit where they could see their own content on display. Visitors who were involved in this part of the project talk fondly about their experience and they also report staying updated and in touch throughout the process. Even though the amount of people who can be involved in this way is very limited, it is a unique and new way of connecting with one's audience. The audience participation was highly praised by one of the museum employees who worked closely in the process:

> I think that (the process) contributes because we develop the audience to be active, not to be just passing visitors who receive some of our stories, and not interfere with them. But to motivate them to be active, to

communicate, to have their own attitude, their own impression, emotions and way of thinking.

The other visitor experience is coming into the museum and seeing the items that others have donated. Upon purchasing a ticket, every visitor is informed about *Your Stories*, as they receive a flyer that describes the process of setting it up and the map of where the extra items are within the exhibit space. To access the experience, visitors need to download the Artcodes application (which is not provided by the museum), but accessible through the Apple Store and Google Play Store on their respective devices.

Many visitors remarked that encountering contemporary objects and personal stories made them reflect differently than they usually would in a history museum. The visit became more emotional, more vivid and for many of the visitors we interviewed it shed a new light on the role of museums in the modern world. Bringing old stories together with modern ones triggered reflection on the human character of history. As our interviewees remarked, this allowed them to see the usually unseen personal aspect of everyday objects. History can easily become depersonalized, when (as in this case) some of the museum artefacts are as old as Stone Age, but pairing them up with contemporary personal accounts, brought them closer to regular everyday experience. There were instances when the museum's pairing did not achieve this effect: Some interviewees felt that the connection between the physical items on display and the digital stories was too vague, and while the stories were interesting they did not see their relevance for the physical exhibit. This seemed to relate to personal preferences in museum spectatorship and understanding this comment fully would require more study. It suffices to notice that although the pairing worked for some visitors it did not work for all, and that the exhibit potentially could be further polished, adapted to the audiences and visitor types that are typical for the museum, on which the museum has more detailed knowledge.

Museum Perspective

Your stories connects to the mission statement of the National Museum of Serbia. The representatives spoke of the importance of activating their audience and helping people to relate to historical content in new meaningful ways. Furthermore, the hybrid character of the installation goes side by side with the museum's policy on using digital technology to find new modes of interactivity. This close integration was made possible through the way the museum was an active participant in the design and development process from day one. Working tightly with the developers as well as reaching out to the existing audience made the museum a central node within the project. The work was not just ordered or outsourced, it was created in-house by incorporating external tech experts in the traditionally museum-contained exhibit development process. This approach allowed the museum greater control and ownership over the project. It also allowed the museum staff to get easy access to guidance in the field of digital solutions and modern technology which often can be very daunting. The synergy between knowledgeable museum personnel and a flexible, technology savvy developer was one of the key ingredients in the success of this project.

Developing the experience came with a very heavy workload for the first 7 months. The process of working with visitor content, matching it with existing exhibits and creating an interactive application to access it all was an immense effort on behalf of the museum and the developer. Creating an experience like this does require a lot of extra research.

Your Stories also allowed the museum to access a new type of content that could then be incorporated into the exhibit as well as to document existing stories pertaining to objects. That is yet another important branch of museum activity that focuses on preserving the current events for posteriority. The stories donated to the museum were often very personal and spoke of current life in Serbia, and as such are of great value to the museum. Getting access to these kind of individual accounts of history can be very difficult so this could be also considered an interesting method for the museum to acquire more content and contextualization within their collection.

Once installed, *Your Stories* requires very little maintenance in terms of staff and online updates. The 3D content and stories are hosted on the museum's own website, making it an integral part of day to day operations of the museum. The information part is done through use of flyers meaning extra personnel is not necessary (although our study did show that extra team members are an advantage). In addition, there are a number of small tasks that are necessary for maintenance of the experience that can easily be forgotten. An important example is the fact that flyers can run out and then someone has to reorder them. Responsibility for that action has been unclear and has caused pauses to the availability of the experience.

Museums have been strongly pursuing the notion of a 'transformative experience' where we understand that a visit at a museum is not supposed to be just a walk among things but it is supposed to stir up something within a visitor, make them think and leave with some new reflection.[3] *Your Stories* was successful in making people think, mostly through the ambiguous coupling with the physical installations. Visitors would focus on trying to figure out the connection between the items and in that process, consider history in two time frames, the *then* and the *now*. *Your Stories* also raises awareness of the fact that all human beings create history throughout their lives, which is another way of increasing visitor reflectivity.

The National Museum of Serbia has been working with a range of new ways of exhibiting their collections, such as for example their 'a museum in a suitcase', a project in which curators would bring interesting items to people physically unable to visit the museum. In that project, focus was on triggering reflection among the audience, which is one of the central concepts within the museum's mission statement. As mentioned in the visitor section, *Your Stories* fosters a more active attitude from visitors, and offers a new layer of engagement that the museum has been looking for. Thanks to the hybrid character of the exhibit, it also works well within the museum space, and doesn't disturb the visitors who do not want to participate while offering a large amount of content to those who are interested. This widens the appeal and variety on offer at the museum, while at the same time it works with content that is deeply meaningful.

Designer Perspective

From the perspective of the designers, the project became a source of new ideas of how to make use of technology within the museum context. 3D scanning has been used for a while in the museum domain, as a way of preserving more fragile historical artefacts. Using 3D scanning to introduce new, less tangible content in to the museum exhibit was a novel approach, that enabled for more content with no requirements on physical space. Working with the artcode technology also allowed for solutions that were interactive, and adaptable on the spot as well as aesthetically pleasing. The combination of the two technologies created a novel approach that the developer considered relevant for design practices more generally, providing opportunities for future explorations of ways to make use of technology in the museum domain.

The option of working with a museum and its audience gave the designers access to content they could normally never interact with and it also allowed

3 Soren, 'Museum Experiences that Change Visitors'.

for an exchange of expertise that is necessary in multidisciplinary projects. Although experienced with technological solutions and interactivity, the knowledge the designers possess on the content of the domain they are working on is limited. Museum staff are an incredible resource as they preserve and are willing to share their knowledge of history and objects, and it's in their interest to find a way to negotiate a good way of fitting the information into new digital channels of distribution. That said, that process is far from an easy one. It requires a certain openness of mind, transparency in dialogue as well as willingness to adapt to the other partner. The positive outcome of this case relied heavily on the tight relationship between the Museum and NextGame. The collaboration resembled that of an in-house tech expert and not an external developer filling a brief for a project. The importance of that personal and mutually respectful relationship between the partners cannot be overstated.

For the designers, it is also very valuable to get to work with real end users/visitors. The people donating the objects and telling the stories offered unique insights into how information could be processed and rendered meaningful. The project illustrates how visitor participation can enrich the design process and give the creators fresh, unexpected ideas as well as push them out of their comfort zone; having the direct contact with the place where the eventual technical solution will be implemented is of great methodological value.

Working this closely with a museum also seems to be a path towards establishing a more prolonged business relationship. Since the development was not done per order but as a shared effort, social connections were established and the collaborators became regular work colleagues on a day to day basis. This reinforced the satisfaction of museum customers and helped the developers retain their presence at the museum.

Analysis of *Your Stories*

Overall, the *Your Stories* experience was positive for all the parties involved. The visitors got a sense of contribution and belonging, as well as new appreciation for historical objects in relation to the present. The museums received support from the designers to deal with technical solutions they were not familiar with, while developers got the necessary domain knowledge that allowed them to customize their design to the particular context. The entire process shared by all the stakeholders was a major reason for why this specific installation was successful. Co-creation polished the raw ideas of all participants into forms that were acceptable and pleasing for everyone involved, introducing new content and new ideas, while also empowering the audiences and fulfilling the mission statement of the museum.

Your Stories also highlights one very important characteristic of tech development for the museum sector. It is never easy and it is never quick, if the goal is for it to be meaningful. The months of iterating, the months of content gathering, multiple versions of prototypes, the production of fliers and hosting solutions all contributed to the final form of *Your Stories*. The experience was precisely adapted to the important stakeholders, it was not rushed and it benefitted from the personal investment of those involved in its development. While the solution is sufficiently generic that it can be implemented in other contexts, the precise process of adaptation was a vital element to its success and at the core of that process lies the tight collaboration between the museums, the developers and the audience.

Bibliography

Barbara J. Soren, 'Museum Experiences that Change Visitors', *Museum Management and Curatorship*, 24.3 (2009), 233–251 <https://doi.org/10.1080/09647770903073060>

Steven Taylor, 'Leaving the Field: Research, Relationships, and Responsibilities', in *Experiencing Fieldwork: An Inside View of Qualitative Research*, 124 vols. (Thousand Oaks: SAGE Publications, Inc., 1991), pp. 238–247 <https://doi.org/10.4135/9781483325514>

About the Author

Paulina Rajkowska is a PhD Candidate at the Department of Informatics and Media at Uppsala University. She comes from a mixed background of sociology, media studies and HCI. Her prior work includes writing on social phenomena in video games as well as application of critical theory to design practices. In her ongoing research she is looking at reflexive practice, playfulness and how both of those can be applied in projects, to improve the quality of their outcomes.

Craft

6. Action Research as a Method for Reflective Practice in Museums

Christian Hviid Mortensen, Anne Rørbæk Olesen, Sejul Malde and Anders Sundnes Løvlie

Abstract

In this chapter we propose action research as a vehicle for reflective practice and organisational learning. We consider organisational learning essential to getting the most out of the digital opportunities and challenges that museums are currently facing. We show how the GIFT action research process functioned as a cross-organisational community of practice (COP) with participants from 10 museums in Europe and USA. The COP provided a safe space for reflection on museum practice for museum professionals and resulted in a set of recommendations on novel ways of working. These ways of working are illustrated with examples from our process. Finally, we discuss some barriers for engaging in action research, embedding learning and sustaining COPs.

Keywords: Action research; Communities of practice; Reflective practice; Organisational change; Innovation; Museum Experiences

Creating hybrid museum experiences such as those presented in this book is no trivial task. Even if many of these designs only require fairly simple technological setups, designing experiences that are both engaging, meaningful and usable requires specialized competences that museums usually lack internally, and therefore need to engage external contractors. However, external help does not entirely remove the need for skills internally. In order to get good results, it is essential that the final design is well aligned with the museum's goals, its mission and its physical environment – including the exhibitions. This requires extensive involvement of the museum's own staff in several stages of the process – ideally including both initial

---

Waern, A. and A. Sundnes Løvlie (eds.), *Hybrid Museum Experiences: Theory and Design.* Taylor & Francis Group, 2022
DOI 10.5117/9789463726443_CH06

research, ideation, and an iterative process going through repeated cycles of prototyping, testing, evaluating and redesigning. How can museum staff – including those who are not IT specialists – gain the necessary skills to participate in such a process? This is not a one-off effort – as technologies and platforms rapidly develop, a continuous effort is needed in order to keep skills up to date.

Donald Schön, a pioneer in the field of organizational learning, has suggested that in response to continuous processes of transformation 'we must invent and develop institutions which are 'learning systems', that is, systems capable of bringing about their own continuing transformation'.[1] But organizational change is notoriously difficult. This is especially true for legacy organizations such as museums that 'are by nature conservative and resistant to change'.[2] Therefore, we argue that we need to get better at reflecting systematically on change at and across museums. Yet, currently in museums there are often only small siloed pockets of internal reflective practice emerging within exclusive communities of practice (COP), divided along narrowly defined professional fields, e.g. conservation, curation or education.

This silo effect is not unique to museums but a prevalent feature of modern organizations.[3] Silos are great for division of labour and specialization and they foster a high degree of accountability. However, this accountability can also lead to internal rivalries and competition for scarce resources. Further, silos can lead to tunnel vision, when we fail to see the full picture of the challenges that are facing us. Silos are not just an organizational feature but also a feature of our minds that structure our way of seeing the world around us. However, we should be aware of our silos and work to counteract their negative aspects.

In this chapter, we outline an approach for museums to engage staff in working collaboratively with digital and hybrid museum experiences through a process which aims to foster internal reflective practice as an essential enabler of change within museums, facilitated by people in all professional roles across the museum. We present action research as a way to accomplish this and disassemble siloed ways of working. Through reflection, experimentation and evaluation, we show how this method enables individuals to become *reflective practitioners*, teams to become

1 Schön, *Beyond the Stable State*, p. 28.

2 Ames, 'Thirty-One Propositions on Changing Museums: An Introduction to the Glenbow Case Study', p. 5.

3 Tett, *The Silo Effect*.

communities of practice (COPs) and, in time, the entire museum to become a *learning organization*. This concept, as defined by Peter Senge, describes an organization which facilitates and encourages collaborative problem solving, continuous learning and transformation.[4] This is an ideal to strive towards – and the benefits come more from striving than arriving. The important part is to maintain a focus on learning. Therefore, we suggest using the learning organization as a guiding metaphor. Rather than trying to change your organization from one state to another, view your organization as in a perpetual state of becoming.[5]

The chapter first introduces the problematic lack of an internal reflective practice culture in museums, then presents how action research can stimulate this kind of culture and how we succeeded in doing so in the GIFT Action Research – an action research project with ten museums from the EU and USA. Finally, the chapter discusses some barriers for engaging in action research, embedding learning and sustaining COPs.

Reflective Practice in Museums

According to Donald Schön a professional practitioner is 'a specialist who encounters certain types of situations again and again'.[6] Following this definition, we find a diverse range of practitioners within museum organizations: From curators to collection managers and digital engagement officers. Through their repetition, practitioners develop a repertoire of expectations and techniques for handling the common tasks they are confronted with in their daily work. This *knowing-in-practice* tends to become increasingly tacit and specialized as the practitioner handles similar cases. While this tacit know-how is displayed in skilful action, a practitioner can foreground it through deliberate reflection, by questioning their current assumptions and understandings of the matter at hand. Schön suggests that this reflection can take the form of a *frame experiment* where a given problem is reframed and imposed on a situation.[7] In this sense reflection-in-action is an exercise in imagination.[8] The reframing transforms the action from knowing-in-action to *reflection-in-action*, where the practitioner attempts to adapt to a new

4 Senge, *The Fifth Discipline: The Art and Practice of the Learning Organization*.

5 Tsoukas and Chia, 'On Organizational Becoming'.

6 Schön, *The Reflective Practitioner: How Professionals Think in Action*, p. 60.

7 Schön, *The Reflective Practitioner: How Professionals Think in Action*, p. 63.

8 Wenger, *Communities of Practice*, p. 217.

situation by providing a solution. Systematic evaluation, as in action research, adds another layer of what we would call *reflection-on-practice*. This can enable failure to be part of innovation by learning from previous mistakes. The aim is not just to do things better, but also to do things differently.

Through our numerous practical projects engaging with museum professionals, we have repeatedly encountered concerns and frustrations among these practitioners that indicate that many museum organizations lack a culture of internal reflective practice. This is possibly related to traditional perspectives about the role of a museum in regard to knowledge and expertise. There exists a hierarchy of epistemology, prevalent in many museums, which prioritizes traditional forms of institutional oriented scholarly research, over other forms of knowledge creation. Recent discourses about pursuing reflective practice in museums are framed explicitly around challenging this knowledge hierarchy.

One of these discourses challenges the assumption that scholarly research is more valuable than other types of research within the museum. This advocates for a greater emphasis on practice oriented research, promoting practical forms of knowledge and focusing on building reflective practice within specific roles, usually within learning or curatorial teams. For example, Ash notes how a recent resurgence in museum educator professional development has focused on reflective practice.[9] Whilst Pringle's work as Head of Research for the Tate has centred on how gallery education practice can be reflectively understood and experienced as a research-led activity, and how learning and exhibition curators can be understood as practitioner researchers.[10]

An alternative discourse challenges the assumption that knowledge generated within the museum is more important than knowledge from external sources, advocating instead for more democratic and participatory forms of knowledge creation involving a wider set of stakeholders external to the museum.[11] In this context, commentators such as Lynch advocate for reflective practice as an essential enabler for museums who are truly committed to working in equal partnership with their communities and building knowledge through citizen science.[12] Several other studies have explored action research in a museum context.[13]

9 Ash, 'Reflective Practice in Action Research: Moving Beyond the "Standard Model"'.

10 Pringle, 'Developing the Practitioner-Researcher Within the Art Museum'.

11 Hooper-Greenhill, 'Changing Values in the Art Museum'.

12 Lynch, 'Custom-made Reflective Practice: Can Museums Realise their Capabilities in Helping Others Realise Theirs?'

13 Stuedahl, 'Participation in Design and Changing Practices of Museum Development'; Tzibazi, 'Participatory Action Research with Young People in Museums'.

While these discourses are welcome advances in developing reflective practice, they arguably still fall short in developing a museum-wide, internal culture of reflection. The latter, whilst advocating for reflective practice to be a core component of all museums, focuses more on the relationships museums have with others, rather than on existing museum work. In contrast the former discourse, whilst focusing specifically on museum work, is limited to specific silos of established practice and seldom addresses the challenges that face the museum as a whole. It can be difficult to develop your practice by acting alone, however, as accurate self-assessment can be complicated. *Deliberate effort* is considered a key factor that separates expert practitioners from the norm.[14] It is better to improve your skills by focusing on *how* you practice, which requires reflection, than just performing a given skill many times. In addition, it requires *qualified feedback*, which can be hard to gain on your own. It also requires *discipline* and *motivation* as the practice is seldom intrinsically rewarding.

These things are easier to sustain in a sociocultural setting and therefore we suggest *communities of practice* (COP) as a suitable vehicle for sustaining reflective practice in, and even beyond, your organization. Lave and Wenger define a COP as:

> [A]n activity system about which participants share understandings concerning what they are doing and what that means in their lives and for their community. Thus, they are united both in action and in the meaning that action has, both for themselves and for the larger collective.[15]

That a practice is a defining property of a community entails that the practice acts as a source of coherence due to a mutual engagement, joint enterprise and shared repertoire of the participants.[16] Further, Wenger argues that a well-functioning COP is particularly suited to 'explore radically new insights without becoming fools [...] A history of mutual engagement around a joint enterprise is an ideal context for this kind of leading-edge learning'.[17]

Figure 6.1 presents a schematic model of the reach and level of reflection from the singular reflective practitioner to COPs with different degrees of

14 Ericsson, Krampe, and Tesch-Römer, 'The Role of Deliberate Practice in the Acquisition of Expert Performance'.

15 Lave and Wenger, *Situated Learning: Legitimate Peripheral Participation*, p. 98.

16 Wenger, *Communities of Practice*, p. 73.

17 Wenger, *Communities of Practice*, p. 214.

Reach		Level of reflection
Individual		Reflective practitioner
Intra-organization	Siloed	Siloed COP
	Across departments	Cross-departmental COP
Extra-organization	Across museums	Cross-organisational COP
	Outside museum sector	Cross-sectorial COP

Figure 6.1: Reach and level of reflection for the different Communities of Practice (COP).

reach within and beyond the organization. We do not consider the level of reflection as normative in the sense that a cross-organizational COP is necessarily better than a cross-departmental COP. Too many participants can turn the COP into an unwieldy entity. However, we encourage you to be aware of the boundaries of your COP and consider the potential value of being more inclusive.

What is Action Research?

We suggest action research as a method for strengthening internal reflective practice in museums. A fundamental tenet of action research is the notion that human systems can only be understood and changed if we involve the members of the system in the inquiry process itself. It builds on a respect for local knowledge and a belief in people's ability to understand and address the issues confronting their community if provided with the proper support and resources. It is participatory, conducted by people who want to do something to improve their own situation.[18]

In its classic form, action research consists of a series of framing experiments, where each experiment is a cycle of five phases.[19] First, we have a diagnostic phase, where a problem is scoped. Then follows an action-planning phase, where the problem is reframed and an alternative course of action or experiment is mapped out. Then the actual action-taking as the experiment is carried out. The experiment is evaluated based on a specified form of data collection in order to establish external validation. Finally, the insights from the experiment is specified by identifying key learnings and preferably documented in a form easily communicated to others.

18 Sagor, *How to Conduct Collaborative Action Research.*

19 Susman and Evered, 'An Assessment of the Scientific Merits of Action Research'.

We consider the key characteristics of action research to be the following:

- Action research is problem-solving or change-oriented. The goal of taking action is altering the status quo in a particular way. By trying to change the situation we gain a better understanding of it.
- Action research is experimental and experiential.
- Action research is iterative.
- Action research is evaluative based on data-collection.
- Action research should document and communicate key findings.

Action Research in GIFT

The action research process that we have engaged in involved participants from ten museums from the USA and Europe: Three art museums, four cultural history museums and three museums with a mixed remit. The process focused on digital design and organizational change. It was structured around five two-day workshops distributed over approximately one and a half years with time for conducting local experiments in between.

Our participants were each asked to form a local working group, ideally consisting of representatives from different departments to form a cross-departmental COP (see Figure 6.2), in which they designed and conducted an experiment with two iterations. Later they performed a second experiment of a more organizational nature with the aim of embedding knowledge gained in the process within the organization.

Many of the participants succeeded in forming a cross-departmental COP. However, some did not, either because of organizational instability, lack of resources in other departments to take part or because the organization was at a stage, or had a size, where the participants found it more useful to start having the discussions internally in one department. Already after the first meeting with the local working group, the participants forming a cross-departmental COP reported on the value of cross-departmental reflection. For instance, one participant became aware of how his work could tie in with the work of a colleague on a different project. Another participant mentioned how differently the cross-departmental representatives understood things and that reflecting on them across backgrounds and orientations were fruitful for unfolding possibilities and developing shared ways of understanding. A third participant reported that:

> They [the other members in the working group] were really more positive than me because I'm so depressed about the whole situation. I actually

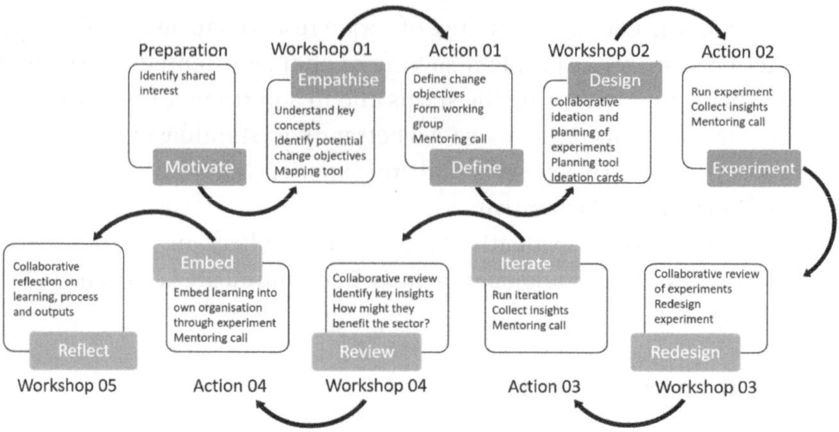

Figure 6.2: Model of the Action Research Process as it unfolded in the GIFT project.

> overlooked a couple of things that we're actually doing–which was great. And I was also impressed by the level of awareness in some points, but in other points, I was kind of surprised about their non-awareness.

This participant also ended up concluding that the institution lacked 'a good cross-departmental infrastructure', something they worked with throughout the action research process. To go deeper into such developments, the next section presents three examples on how the cross-departmental COPs helped push the participating institutions towards becoming learning organizations.

The Cross-Departmental COP: Three Examples

For the second, organizational experiment, the Munch Museum in Oslo initially ran a workshop with participants from across their organization focused on how to successfully employ their digital collections. However, the enthusiasm and productive atmosphere experienced by participants in the workshop formed the impetus to immediately organize another workshop with more open discussions on how to potentially transform the organization. In fact,

> [T]he form, process and atmosphere of the workshops were in themselves cases in point for how we want to work at the museum in the future. That is, in small, agile and interdisciplinary project groups that work together in a friendly and productive atmosphere on the basis of mutual respect,

> thereby overcoming silo thinking within the confines of department structures.[20]

The creative climate of the initial workshop fuelled an ongoing process where the group became a digital think tank that met on a regular basis to talk about the digital transformation of the Munch Museum. At a later stage – after the end of the GIFT project – the work in this group led to the establishment of an Audience Lab in the museum, now a permanent part of the organisation with an aim to drive innovation in digital visitor experiences at the museum.

The Munch story is a case of a small-scale experiment, establishing a cross-departmental COP around digital collections, resulting in durable changes in workflows and with potential future ramifications for the entire organization. Another of our museum partners, The Norwegian Centre for Holocaust and Minority Studies, embarked on a more radical course. After experimenting with the concept of playfulness as a novel approach to the complex, emotionally charged and sensitive subject matter of the centre as part of the GIFT Action Research, they decided to continue 'playing around' as a way to develop a new strategy for the upcoming extension of their gallery space in a more inclusive manner.[21] The ambitious aim was to create a more democratic culture within the organization. Through a series of three workshops they established a space for cross-departmental collaboration that included everyone from front desk staff to management. Before the events the organizers encountered scepticism from colleagues and management that were less enthusiastic about democratizing the organization. During the workshops, however, scepticism was transformed to engagement when they discovered the creative enthusiasm found in all levels of the organization. Each workshop contained a visit to another institution providing an outsider perspective and opportunity to share professional experiences and in this way extending the COP beyond the organization. On the surface the experiment succeeded in including everyone from the organization. However, feedback from the participants revealed that colleagues from marketing, front desk and maintenance found it difficult to contribute as much as they wanted to. This might be because they were less accustomed to open ideation processes. This offers a learning point for future projects: When involving professionals from different backgrounds in such ideation processes one should make sure that the tasks and procedures involved are

20 Mathias, 'From Website Discussions to Transforming the Organisation'.

21 Christensen, 'Working Across Departments to Find Playful and Inclusive Strategies'.

made sufficiently clear that everyone may contribute fully, regardless of prior experience. Also, it became apparent that written feedback was essential, as not everyone is comfortable voicing their honest opinions publicly in front of colleagues and management. Therefore, they urged participants to compile 'exit notes'. As it turned out, the opinions expressed in these notes often differed from the ones voiced during the workshop.

The experiments at the Munch Museum and The Norwegian Centre for Holocaust and Minority Studies were successful in stimulating new organizational initiatives and setting the stage for organizational change. A common denominator for the two case stories is that each organization was in the midst of major redevelopments that offered a clear, shared purpose. The Munch Museum was relocating to a new purpose built museum building, while the Holocaust centre was building a sizable extension to their existing exhibition space. In both cases, these major redevelopments created favourable conditions for rethinking organizational orthodoxy. Also, the challenges involved encouraged everyone to work together toward a common goal.

It can be harder to ascertain a clear purpose and reason for disrupting business-as-usual, when not faced with such immediate challenges. This might have been a factor in another experiment, that failed to get fully off the ground. Tyne & Wear Archives & Museums wanted to instigate a culture of productive failure: '[A] culture where the "imperfect" and the "unfinished" are tested with the public in a cycle where products and projects are constantly being improved as a result of feedback.'[22] Central to this approach is the belief that failures offer crucial opportunities for learning and therefore that the fear of making mistakes can impede innovation. In order to do this, they wanted to celebrate failure at a social event, where practitioners could share stories of their professional failures and thereby collectively learn from the mistakes of others. At first, the event was envisioned to be a safe space open for all professionals in the cultural sector, however it was quickly scaled down to be for staff only, before being dropped altogether. The notion of openly admitting failures proved to be too sensitive and made people nervous. Instead, the organizers decided to start on a smaller scale by running an experimental prototyping session with staff. The session helped engage staff in the theme of failure in a 'safer' way – in that if something does not work, you need to tweak and tinker. 'Failing forward' in this way proved easier than rehashing old failures.[23] This example of a cross-departmental

22 Younas, 'From Failure Cafés to Rapid Prototyping'.

23 Maxwell, *Failing Forward: How to Make the Most of Your Mistakes.*

COP at work shows us that changing people's perception and work practices takes time. Especially, when there is no imposing challenge establishing a common purpose. On the other hand, one could argue that this is in fact an opportune moment in which to revisit failures of the past, when the organization is not pressed by immediate challenges and can therefore allocate the time and space for such reflections.

All three case stories display the workings of a cross-departmental COP in action within an organization. However, our museum partners also spoke of the value of partaking in a COP beyond their own organization.

The Cross-Organizational COP: Learning Beyond the Organization

The group of museum professionals participating in the GIFT project became an international and cross-organizational COP. The participants acted together during the workshops in a concerted effort to develop their own practice through experimentation. These actions established a joint enterprise, which was negotiated and defined by the participants in the process of pursuing this enterprise. Through the process, participants developed a shared discursive repertoire by which to express and talk about issues.

Participating in the cross-organizational action research process provided participants with the opportunity to act on and think about the challenges they had encountered in their daily practice without organizational constraints such as accountability and hierarchical power relations. The cross-organizational COP can act as a *safe space* for unorthodox thinking, which can generate a lot of creativity. However, this energy is often drained when participants return to the political milieu of organizational life fraught with power relations.[24]

The notion of the cross-organizational COP as a safe space is supported by the feedback we received from our participants. The feedback centred around four themes:

- They valued that the process had been slow, allowing time for reflection and resulting in a relaxed atmosphere.
- They valued the openness and honesty of our discussions. They enjoyed the workshops as a space free from the normal routines, constraints and power relations where they could share ideas and experiences.

24 Pemberton, Mavin, and Stalker, 'Scratching Beneath the Surface of Communities of (Mal) Practice'.

- They valued our conversations for the learning and reflection that they engendered.
- The valued the international collaboration with museum professionals with diverse cultural backgrounds, resulting in multiple perspectives on common issues.

The GIFT Project: A Cross-Sectorial COP

The action research in GIFT was integrated within the overall research project in a way that turned the various participants in GIFT into a cross-sectorial COP consisting of university researchers, artists, designers, as well as the museum professionals participating in the action research. This broader community was formed partly as a result of research partners contributing to the work of the action research participants – by participating in the workshops as speakers and facilitators, as well as by entering into collaboration with the participants. For example, Blast Theory initiated a collaboration with Brighton Museum in the development of the *Gift* app (presented in Chapter 3). Furthermore, there were also collaborations which were initiated by the action research participants, who took an interest in the research done in other parts of GIFT. Notably, the Munch Museum took on collaborations with both Blast Theory and NextGame (the company behind the *Your Stories* project presented in Chapter 5). Brighton Museum got involved in the development of a guide app at the IT University of Copenhagen called *One Minute* – a project which was later taken over by the museum and is still under development with the aim to become part of the museum's digital services.[25] The action research participants also on several occasions contributed to the larger research project by testing and giving feedback on designs and prototypes – for instance, two of the sensitizing scenarios presented in Chapter 7 were tested in one of the action research workshops. One of the action research experiments conducted by the museum participants even inspired another research project on image recognition within artworks at the IT University of Copenhagen, a project which is still running at the time of writing.[26]

An important catalyst for these cross-sectorial collaborations was the decision to colocate each of the first four action research workshops

25 Løvlie and others, 'Designing for Interpersonal Museum Experiences'.

26 Kadish, Risi, and Løvlie, 'Improving Object Detection in Art Images Using Only Style Transfer'.

with a project meeting for the rest of the GIFT participants, at one of the participating museums. This not only enabled the other participants in the research project to contribute to the workshops; it also ensured that participants from all parts of GIFT would meet and mingle repeatedly over a period of eighteen months, providing sufficient time and opportunities for formal and informal networking that allowed participants to get to know each other, and discover common interests and opportunities for collaboration.

However, establishing a COP with such different participants was not without challenges. First, the conditions for participating differed quite significantly for different participants. For many of the university participants, GIFT was either a full-time commitment or a significant part of their workload for the three years that the project lasted, whereas the participants in the action research process had much more limited time dedicated to work on the project in the action-taking phases between the workshops. This asymmetry sometimes made it difficult for the action research participants to engage on an equal footing with the other project partners.

Furthermore, it was challenging to create a high degree of alignment between the themes of the action research process and the themes that were in focus of the larger research process. Action research starts with the goals and the challenges of the participants. In our context, this meant that a significant part of the action research process was dedicated to exploring the goals and challenges the participants were encountering in their respective institutions, and developing a shared focus for the group. This led the action research group to have a focus that was somewhat broader in scope than the rest of GIFT, as a result of exploring the many issues – organizational as well as technical, political, educational, economical, etc. – that influenced their organization's ability to innovate. This reflects a common tension with practice-based research in an academic context: Academic research tends to require a high, or even extreme degree of specialized focus on a narrowly defined research question grounded in theoretical literature – whereas practical design and organizational work often needs to address a broader variety of issues that are relevant to the challenge at hand. A key learning for future projects that bring museum professionals in collaboration with researchers and designers in a similar way would be to invest significant effort into creating strong alignment between the work processes of researchers and practitioners from different sectors. Creating rich arenas for both formal and informal connections between participants may form a valuable part of such an effort.

Discussion

The cross-departmental COPs and cross-organizational COP formed in connection with the aforementioned action research process were clearly valuable. However, action research can only happen within one's sphere of influence. Therefore, if action and change requires collaborative support and the combined effort of several teams or even the entire organization this becomes increasingly difficult as organizations are frequently gridlocked – caught between the forces of change and the current ways of doing things. Discussions between these fractions are often futile as they are based on assumptions and biases. Action research can help resolve such a gridlock by providing new data.[27] A key objective of action research is to produce data that can substantiate any claims made. Then discussions can become about the data and not our biases. Evidence that calls our previous behaviour into question can be an agent of change that causes us to reconsider this behaviour. However, if information, or data, does not build up to participation – when 'it fails to translate into a way of being in the world coherent enough to be enacted in practice' – it remains alien, abstract and fragmented.[28] This enacting and participation is what situates learning and transforms abstract information into practical knowledge. This was the case with the failure café at Tyne & Wear, for example, where the concept of failure was too abstract and alien to people in the organization, resulting in the decision to focus on experimental prototyping as a smaller step towards changing people's perceptions and practices.

This also touches upon the problem of embedding the learning gained via action research in the organization. How can this knowledge be transferred from the COP where it was developed to the rest of the organization? The organizational literature on embedding learning or innovations in organizations often centres on the role of *innovation champions* as those actors who take an innovation on board and adapts it to fit the organizational context, not to be confused with *innovators* who are the actors who developed the idea in the first place.[29] Innovations need champions in order to survive in an organization. Championing entails identifying, refining and supporting the innovations that are introduced.[30]

27 Sagor, *How to Conduct Collaborative Action Research.*
28 Wenger, *Communities of Practice*, p. 220.
29 Rogers, *Diffussion of Innovations.*
30 Zahra and others, 'Fostering Entrepreneurship During International Expansion'.

There is an element of risk involved in championing innovation, as it is by definition unknown territory and any adoption will require spending resources which are often scarce. Thus, the champion needs to advocate their case for the added value of a given innovation to convince any sceptics and for colleagues to rally behind the idea.[31] Successful innovation champions often occupy senior positions, as it requires a certain clout to enforce innovations within organizations and give them enough legitimacy to become institutionalized, as was the case at the Munch Museum.[32] Therefore, senior personnel can be either a *driver*, a *blocker* or *neutral* towards innovations. Further, the role of the innovation champion does not rest on the innate qualities of the individual. Rather it is a socially constructed identity that is continuously re-negotiated within the organization.[33]

Similarly, COPs can be fluid, changing and loosely constructed spaces for reflection. Engaging a COP in action research, however, formalizes reflection, resulting in actions and potential changes taking place in the planned period of the action research. But what happens when the period ends? How is the experimental and reflective space to be sustained? While the action research is valuable in itself, it is worth thinking about the potential value of continuing the COP in a less formalized way. Or a more formalized way, as in the Munch Museum case where the museum turned the informal digital think tank into a permanent part of the organizational model (now named the Audience Lab). Thus, formalizing the group within the organizational structure of the museum might add more leverage, but also risks making the group exclusive and self-contained, which can prevent fresh perspectives from outsiders, something which they explicitly valued in the initial workshops. Thus, thinking about the sustainability of your COP and how it 'lives' within your organization is essential: What is needed in order for your COP to prosper, rather than perish?

Conclusions

In this chapter, we have argued for the need to get better at reflecting on change at and across museums. We have advocated that organizational learning processes can start with the individual reflecting on their own practice,

31 Jenssen and Jørgensen, 'How Do Corporate Champions Promote Innovations?'

32 Garud, Tuertscher, and Van de Ven, 'Perspectives on Innovation Processes'; Van de Ven and others, *The Innovation Journey*.

33 Sergeeva, 'What Makes an "Innovation Champion"?'

preferably in a systematic way, such as action research. Further, our goal has been to show how different levels of COPs can be a vehicle for reflective practice and, in time, organizational learning. Despite their apparent value, we have highlighted the scarcity of forming cross-departmental COPs with participants from other departments and cross-organizational COPs with participants from other museums. Through the action research process of the GIFT Project, we have shown examples of the work and value of cross-departmental COPs and cross-organizational COPs, as well as our experiences with the overall GIFT project as a cross-sectorial COP. Lastly, we have discussed barriers for engaging in action research, embedding learning and sustaining COPs.

Importantly, our aim has not been to say that one kind of COP is better than another. On the contrary, they provide different levels of reflection, as highlighted in Figure 6.1. Even though we have focused on cross-departmental COPs and cross-organizational COPs, we acknowledge the value of other kinds of COPs. For instance, we can envision the value of different types of cross-sectorial COPs that include other sectors than those represented in GIFT, e.g. to include members of the local community or practitioners from other fields. Such outside perspectives may help to counteract silo-thinking following organizational and sector structures.

One might argue that an action research process such as the one in the GIFT project is not feasible for smaller organizations without extra resources such as the backing of a large externally funded research project. We believe, however, that the main purpose of action research is to reflect systematically on what you are already doing and engage in small scale experiments with improving your ways of working; something that can be achieved without setting up large resource demanding development projects. The point of departure is always the current practice and good innovation begins with ideation, not expensive technology. We want to conclude by encouraging readers who work in museums, whether in a small or large institution, to reach out to other practitioners within and beyond your organization in order to reflect and learn.

Bibliography

Michael Ames, 'Thirty-one Propositions on Changing Museums: An Introduction to the Glenbow Case Study', in *Ed. RR Janes, Museums and the Paradox of Change* (Calgary, Canada: Routledge, 1997)

Doris Ash, 'Reflective Practice in Action Research: Moving Beyond the "Standard Model"', in *The Reflective Museum Practitioner* (New York: Routledge, 2019), pp. 23–38

Elise Grimsrud Christensen, 'Working Across Departments to Find Playful and Inclusive Strategies', *The GIFT Box*, 2019 <https://gifting.digital/norwegian-center-for-holocaust-and-minority-studies/> [accessed 2 April 2020]

K Anders Ericsson, Ralf T Krampe, and Clemens Tesch-Römer, 'The Role of Deliberate Practice in the Acquisition of Expert Performance', *Psychological Review*, 100.3 (1993), 363–406

Raghu Garud, Philipp Tuertscher, and Andrew H Van de Ven, 'Perspectives on Innovation Processes', *Academy of Management Annals*, 7.1 (2013), 775–819

Eilean Hooper-Greenhill, 'Changing Values in the Art Museum: Rethinking Communication and Learning', *International Journal of Heritage Studies*, 6.1 (2000), 9–31 <https://doi.org/10.1080/135272500363715>

Jan Inge Jenssen, and Geir Jørgensen, 'How Do Corporate Champions Promote Innovations?', *International Journal of Innovation Management*, 8.01 (2004), 63–86 <https://doi.org/10.1142/S1363919604000964>

David Kadish, Sebastian Risi, and Anders Sundnes Løvlie, 'Improving Object Detection in Art Images Using Only Style Transfer', in *The International Joint Conference on Neural Networks (IJCNN)* (presented at the The International Joint Conference on Neural Networks (IJCNN), Virtual Event: IEEE, 2021) <https://arxiv.org/abs/2102.06529> [accessed 23 June 2021]

Jean Lave, and Etienne Wenger, *Situated Learning: Legitimate Peripheral Participation* (Cambridge U.K.: Cambridge University Press, 1991)

Anders Sundnes Løvlie, Lina Eklund, Annika Waern, Karin Ryding, and Paulina Rajkowska, 'Designing for Interpersonal Museum Experiences', in *Museums and the Challenge of Change: Old Institutions in a New World*, ed. by Graham Black (London & New York: Routledge, 2020), pp. 223–238

Bernadette T Lynch, 'Custom-made Reflective Practice: Can Museums Realise their Capabilities in Helping Others Realise Theirs?', *Museum Management and Curatorship*, 26.5 (2011), 441–458

Nikita Mathias, 'From Website Discussions to Transforming the Organisation', *The GIFT Box*, 2019 <https://gifting.digital/munch-museum/> [accessed 4 June 2021]

John C Maxwell, *Failing Forward: How to Make the Most of Your Mistakes* (Thomas Nelson Publishers, 2000)

Jon Pemberton, Sharon Mavin, and Brenda Stalker, 'Scratching beneath the Surface of Communities of (Mal)Practice', *The Learning Organization*, 2007

Emily Pringle, 'Developing the Practitioner-Researcher Within the Art Museum', *Tate Papers*, 29 (2018), 16–17

Everett Rogers, *Diffussion of Innovations*, 5th edition (New York: Simon & Schuster, 2003)

Richard Sagor, *How to Conduct Collaborative Action Research* (Alexandria, VA: Association for Supervision and Curriculum Development, 1993)

Donald A. Schön, *Beyond the Stable State* (New York: Random House, 1971)

Donald A. Schön, *The Reflective Practitioner: How Professionals Think in Action* (Aldershot, UK: Ashgate, 1983)

Peter M Senge, *The Fifth Discipline: The Art and Practice of the Learning Organization*, 2nd edn (New York: Doubleday/Currency, 2006)

Natalya Sergeeva, 'What Makes an "Innovation Champion"?', *European Journal of Innovation Management*, 19.1 (2016), 72–89 <https://doi.org/10.1108/EJIM-06-2014-0065>

Dagny Stuedahl, 'Participation in Design and Changing Practices of Museum Development', in *The Routledge Handbook of Museums, Media and Communication*, ed. by Kirsten Drotner, Vince Dziekan, Ross Parry, and Kim Christian Schrøder, Routledge International Handbooks (London and New York: Routledge, 2019), pp. 218–231

Gerald I. Susman, and Roger D. Evered, 'An Assessment of the Scientific Merits of Action Research', *Administrative Science Quarterly*, 23.4 (1978), 582–603 <https://doi.org/10.2307/2392581>

Gillian Tett, *The Silo Effect: The Peril of Expertise and the Promise of Breaking Down Barriers* (New York: Simon and Schuster, 2015)

Haridimos Tsoukas, and Robert Chia, 'On Organizational Becoming: Rethinking Organizational Change', *Organization Science*, 13.5 (2002), 567–582 <https://doi.org/10.1287/orsc.13.5.567.7810>

Vasiliki Tzibazi, 'Participatory Action Research with Young People in Museums', *Museum Management and Curatorship*, 28.2 (2013), 153–171 <https://doi.org/10.1080/09647775.2013.776800>

Andrew H. Van de Ven, Douglas Polley, Raghu Garud, and Sankaran Venkataraman, *The Innovation Journey* (Oxford, UK: Oxford University Press, 2008)

Etienne Wenger, *Communities of Practice: Learning, Meaning, and Identity*, Learning in Doing: Social, Cognitive and Computational Perspectives (Cambridge: Cambridge University Press, 1998) <https://doi.org/10.1017/CBO9780511803932>

Sarah Younas, 'From Failure Cafés to Rapid Prototyping', *The GIFT Box*, 2019 <https://gifting.digital/tyne-and-wear/> [accessed 2 April 2020]

Shaker Zahra, James Hayton, Jeremy Marcel, and Hugh O'Neill, 'Fostering Entrepreneurship During International Expansion: Managing Key Challenges', *European Management Journal*, 19.4 (2001), 359–369 <https://doi.org/10.1016/S0263-2373(01)00037-8>

About the Authors

Christian Hviid Mortensen, PhD, is a former curator of media heritage at the Media Museum in Denmark (2007-2018). Currently Christian is a postdoc at the IT University of Copenhagen, where he researches the dynamics of hybrid heritage experiences spanning the digital and physical realms. His research interest is the dynamics between media, culture, heritage and memory. Christian is also on the editorial board of MedieKultur: Journal of Media and Communication Research.

Anne Rørbæk Olesen earned her PhD in Communication from Roskilde University in 2015 with the thesis *Co-designing Digital Museum Communication*. She has co-authored publications in internationally renowned anthologies and journals, such as *The Routledge Handbook of Museums, Media and Communication, Museum Management and Curatorship* and *Museum & Society*. Her research focuses on digital communication, design practices, collaboration and experience design.

Sejul Malde is a Cultural Research Manager. He works with cultural organisations, developing and managing collaborative research projects that explore how cultural organisations respond to digital change and become more relevant to the needs of today's audiences and society. Sejul brings significant experience from across both the commercial and cultural sectors. He is currently based in Australia, working as Research Manager for the College of Arts And Social Sciences at the Australian National University.

Anders Sundnes Løvlie does research on the intersection of design research and media studies, focusing in particular on experience design, locative media and play. Anders was the coordinator for the GIFT project, and has been involved in a number of design projects involving museums like the Munch Museum, Danish Architecture Center, the National Gallery of Denmark, Brighton Museum and the Frederiksberg Museums. Author of the book *The Rhetoric of Persuasive Games: Freedom and Discipline in America's Army*. Leads the research group on Media, Art and Design (MAD).

7. Sensitizing Design Teams to Theory

Annika Waern and Paulina Rajkowska

Abstract

This chapter is the first of the practical methods and tools chapters, and addresses a well-known problem in design research: How can relevant insights from academic theory help to inform practical design processes? The chapter details two possible ways in which this can be done: through the explicit incorporation of theorists into the design team, and through sensitising exercises such as role-played scenarios. The chapter shares insights into how these methods work, and present concrete suggestions for how to implement them in practice.

Keywords: Sensitising concepts; High theory; Sensitising designers; Design team; Design processes

One of the challenges that hybrid design projects face is to get everyone involved in a design project on the same page. The participants in such a project will typically have very different backgrounds and competencies; they can be curators, pedagogues, managers, marketing personnel, designers, programmers and artists. They work in varying roles; some will be tasked with design and implementation, some will provide content, others will work with implementing and marketing the service in the museum, and some represent the museum as a stakeholder. They also work under very different conditions, as employees or consultants, for the duration of the project or just briefly. Typically, design projects are also distributed in space and time, with little opportunity for the project group as a whole to meet; yet, everyone needs to be involved and contribute to the project. Developing a mutual and joint understanding of the project, project goals and design approach can thus become very difficult, especially since these are seldom fully articulated when a project starts (this will be further discussed in Chapter 8).

Waern, A. and A. Sundnes Løvlie (eds.), *Hybrid Museum Experiences: Theory and Design.* Taylor & Francis Group, 2022
DOI 10.5117/9789463726443_CH07

A critical resource in the dialogue around design goals and ideals is *language*, but in practice, people with such diverse backgrounds and roles as sketched above will not speak exactly the same professional language. Their expertise will be expressed in terms that not everyone will understand, and even more challenging is when people use similar or the same words to express very different things. This is not an uncommon problem in development projects, but when it comes to hybrid museum experiences this happens all the time. Take the word 'experience' to start with – this is a term that will mean very different things for the museum pedagogue and the programmer. In GIFT, we encountered problems with the term 'appropriation', which had positive connotations for the interaction designers (as a term denoting visitors taking control of their technology), but negative for the curators (as in cultural appropriation).

Language issues are prone to arise when designs aim to tap into social practices. When we talk about things like 'family', 'friendship', 'gifting' and 'play' or even 'museum visitor' we all have an intuitive understanding of what we mean, and by that what we *think* that others mean. But our everyday understanding can very easily lead us astray – all of these are complex practices that vary over time, between cultures, and between individuals. They are words that come with an inherent risk of different project members meaning very different things, but also that unless the complexity of the concepts is made explicit, the group will end up designing for a very shallow understanding of the underlying phenomenon.

One way that such hurdles can be overcome, is through including activities that help to *sensitize* team members towards critical concepts in their design approach, with the goal of eliciting the different possible meanings, enabling the design process to stay focussed on key goals and aware of potential issues. This chapter presents two different ways in which this can be done. *Sensitizing scenarios* are exercises that can be run within the team at suitable times as a way to create group cohesion around complex concepts. The second approach is to include a *resident theorist* in the team. We report on our experiences with both methods and discuss when they may be appropriate. The chapter ends with a brief introduction to alternative methods, if these two should be considered too difficult or time-consuming to adopt.

Sensitizing Concepts and Boundary Objects

Both methods presented in this chapter are based on a discourse-analytical framing of the problem of diversity in design teams. This means that they centre on language and our use of words to communicate, establish common ground, and articulate our differences in opinions and approaches. While language is not the *only* thing a team needs to share in order to create a mutual grounding of a design project, language must be considered a critical resource.

The methods proposed in this chapter are grounded in an analytical perspective on sociological theory, as manifesting in the form of *sensitizing concepts*. This perspective was originally articulated by Blumer.[1] Blumer argued that in sociological theory it makes little sense to construct formal definitions; the phenomena that sociology aims to capture are too complex and the boundaries between them too diffuse. Blumer suggests to instead view central concepts as 'sensitizing': Their role is to sensitize the researchers to phenomena that manifest over and over again, but every time in a unique way. 'A sensitizing concept', Blumer writes, 'gives the user a general sense of reference and guidance in approaching empirical instances. Whereas definitive concepts provide prescriptions of what to see, sensitizing concepts merely suggest directions along which to look'.[2] The methods proposed in this chapter are ways in which sensitizing can be fostered, in other ways than by long-term academic studies or field observations.

The idea of sensitizing concepts has been adopted in design research as a way of articulating design knowledge. In design research, sensitizing is typically seen as an early step in a design process, to be followed by the identification of requirements, design resources and engaging in various design exercises. In the previous chapters, we have shown how abstract theories such as sociological theories around social ties, anthropological studies of gifting, and psychological perspectives on social play all can be relevant for design. More generally, Zimmerman, et al. (2010) suggest that more broadly scoped 'guiding philosophies, which take the form of sensitizing concepts' can work to help direct designers and researchers in solving design problems.

A related concept is that of 'boundary objects'.[3] These are shared structures including physical objects, that allow actors with differing goals to work together through agreeing on shared resources and ways of working. In their original study, Star and Griesemer investigated how amateurs and professionals managed to work together, in acquiring the collection for the Berkley Museum of Vertebrate Zoology. Some examples of the boundary

1 Blumer, 'What is Wrong with Social Theory?'
2 Blumer, 'What is Wrong with Social Theory?', p. 7.
3 Star and Griesemer, 'Institutional Ecology, "Translations" and Boundary Objects'.

objects they identified were the collections as such, various forms to fill in, and clear geographical borders for the area from which the collection was acquired. But they also discuss the importance of *ideal types*, concepts enabling communications across the worlds of the experts and amateurs, e.g. maps and names for species. Later, Star clarified that 'boundary objects are a sort of arrangement that allow different groups to work together without consensus'.[4] They differ from standards in that they arise organically from a perceived need. Words and language are typically not in themselves boundary objects, because a boundary object also needs a process, something to do. In the context of design, sensitizing concepts can be approached as boundary objects: A sufficiently common ground must be established to allow the parties to use them in their work and communication, while leaving room for the parties to ground them further to be meaningful in their own practice in slightly different ways.

Sensitizing Scenarios

There exist obvious ways for teams to establish a common theoretical grounding for a design team – such as having everyone read the same books and discuss them together. Time and resource constraints will however often make this impossible, and if the core concepts are from social theory, the discussions risk becoming rather abstract. In this section, we propose a different option: *Enacting* them in the form of role-play. *Sensitizing scenarios* are short, scripted sessions of role-play, that allow people to engage in a social situation as someone else than themselves, and also reflect on the theoretical grounding for the experience. Our experiences with using sensitizing scenarios as part of a design process within and outside of GIFT have been reported in more detail in a study by Waern and colleagues.[5]

The Role of the Scenarios in a Design Process

In a role-played scenario, participants are instructed to improvise the actions and reactions of a character. Henriksen defines role-play as 'a medium where a person, through immersion into a role and the world of this role, is given the opportunity to participate in, and interact with the contents of this world,

4 Star, 'This is Not a Boundary Object'.

5 Waern and others, 'Sensitizing Scenarios'.

and its participants'.[6] Role-playing is possible in two ways: Players can play themselves in fictional situations, or they can take on fictional characters, with their own attitudes, personalities and goals. The tradition of interactive and improvisational drama has long worked to develop ways to use role-play for community-building, learning and reflection among the participants.[7] What sets this genre of drama apart from theatre is that it is typically not performed for a separate audience, but for the benefit of the participants.

There are several reasons why role-play is an option when it comes to sensitizing in a museum design project context. Firstly, it is fairly time efficient; a scenario can be designed to take less than an hour to run and still be useful as a sensitizing exercise. Secondly, role-playing gives access to a *first-person experience*. It allows participants to immersively engage with a phenomenon and gives a glimpse of a person's bodily and affective responses to it. (This should not be confused with actually living through the same – experiencing something within the safety net of a staged experience is different, and sometimes radically so, from experiencing the same thing in everyday life.) Finally, when role-playing is done by the whole project team together, it forms a *shared* experience that they can later on discuss and refer back to. Experiencing something through bodily and affective engagement creates rich memories and promotes reflection.

An Example Scenario

While role-playing scenarios can take multiple forms, we developed and tested a range of scenarios within the GIFT project, a process that allowed us to establish a common structure and time frame that we believe is suitable for the purpose of sensitizing. The GIFT scenarios include three phases: A preparation phase, a play phase and a concluding debrief during which the relationship to theory is deepened and discussed. They are designed to be playable in small groups and within max 1.5 hours, and they all require a person to act as *facilitator*. The facilitator makes all necessary preparations before players arrive, has read (and preferably also previously played) the scenario, and guides the players through playing and debrief. (The facilitator can participate in the scenario as a player if desired or needed.)

One example is the scenario *The Object*.[8] In this scenario, the participants follow an object from its creation, via the original cultural context and

6 Henriksen, 'Learning By Fiction'.

7 Blatner, *Interactive and Improvisational Drama*.

8 Available in full from https://gifting.digital/the-object-3/ (Accessed March 2020).

1. The emotional You love this quirky object and don´t want to get rid of it.	**6. The rule obeyer** You want to follow the guidelines. And make the right decision based on that.
2. The horder You never want to get rid of anything. Keep everything!	**7. The crowd pleaser** All you care about is what is best for the audience and the public opinion.
3. The pragmatic If it is no longer of use, get rid of it.	**8. The storyteller** Does the object have an interesting story? If not, get rid of it.
4. The planner Knows that the museum needs more space so we need to get rid of stuff	**9. The team player** You just want to find a solution everyone is happy with
5. The effective You just want this to be over quick, make a decision, you don't care so much which.	**The facilitator** Plays an intendent in this scene. Goal: To lead the meeting and get the group to make a decision.

Figure 7.1. Roles for one of the scenes in 'The Object', called 'The Ending'.

until it ends up in a museum, is put on display and finally archived or destroyed. The intended use of the scenario is within design projects that deal with museum collections where the history of objects is contentious. It problematizes how museum collections sometimes have a dark past, and museums often know very little about the objects in their collection, their history and their creation.

As in most role-playing scenarios, it requires one person to act as facilitator. Before players arrive, the facilitator needs to prepare the scenario, for example by choosing an object that will be used as the thematic focus for the scenario, and printing out all the roles. In this particular scenario, the facilitator also takes on a supporting character in each of the scenes.

Figure 7.1 presents some of the roles which players get to play in this particular scenario. It should be noted that their descriptions are very brief, focussing on a short characterization of their emotions and goals. In most scenarios, character descriptions are slightly longer and include, e.g. how they are related to each other. However, for this particular scenario players take on new roles for each scene, making it necessary to minimize their descriptions. Apart from the short character text, players can improvise everything else about their character.

The play phase in *The Object* consists of a sequence of short scenes detailing important events in the object's history. Most of the scenes have set outcomes: The players can argue as they wish based on their roles, given that they eventually reach the set outcome. The scenario also has a set ending: The object is to be culled from the collection and destroyed.

The debrief session is a critical part of a sensitizing scenario. Its purpose is to help players reflect on the experience and re-contextualize their experiential learning process from the fictional setting of the scenario, to problems situated in their design project.[9] In the sensitizing scenarios developed within GIFT, recontextualization is fostered through the prescripted debrief. We have found it useful if the facilitator first introduces the background concepts, theories, and research in a structured way, before opening up for a general discussion and reconnection to the design project at hand. In *The Object*, the debrief consists of two phases. First, the facilitator shortly presents the central concepts underlying the events in each of the scenes (as given by the scenario). Next, players and the facilitator engage in an open discussion. The script contains some suggestions for discussion topics for this concluding discussion.

Staging a Scenario

Within GIFT, we have gathered a range of experiences with staging scenarios with museum professionals and designer teams. In two cases, we were invited to stage scenarios with designer teams from ongoing projects developing hybrid museum experiences. For both of these, the projects had been going on for some time and had experienced difficulties in achieving a joint vision for their design concept.

While the scenarios were originally designed to be staged very early in the design process, this setting had some advantages. Since much was already known about the requirements and assets available for the project, it was easy to select the appropriate sensitizing scenario, and adapt it to the project at hand. However, this late during the design process means the design might have advanced to a stage where it is difficult to make major changes. Even if engaging with the scenario would lead to drastic

9 Henriksen, 'Learning By Fiction'.

changes in the teams' perspectives, it may no longer be possible to let that influence the design.

It should be noted that the reconnection and contextualization process that happens after running a scenario, is intrinsically dependent on there being a context to connect to. For example, when running *The Object* with a museum project team, the participants would immediately connect to people in their own organization ('that person works at marketing!'). When run with a project team we can expect this reconnection and contextualization process to continue also after the debrief. Hence, it is not necessary to close the discussion fully within the debrief – some questions and reflections can be left for later.

A sensitizing scenario will typically create heightened group cohesion and a high level of energy in the group. Since this is likely to fade quickly, it is useful to go directly into a design exercise after the debrief. This can be a requirements elicitation exercise (as discussed in Chapter 8), a brain- or bodystorming exercise, or a preliminary evaluation of a suggested design, depending on how far the project has progressed. We used slightly different exercises in our two project workshops.

Practical Tips and Tricks

Running a sensitizing scenario makes sense under the following conditions:
- There is a good fit between the design project and the chosen role-play scenario. For example, it may be relevant to run a scenario about cultural appropriation before planning an exhibition on exotism.
- Not all members of the project team are well versed in the topic for the scenario.
- In order to create a meaningful design, the project team needs to develop both an intellectual and an affective/emotional connection to the topics of the scenario.

The last point is important. Most likely, the project team will need to get a shared understanding of many different things, including the available budget and the target technology. But for those, there are other methods that are more efficient and appropriate (e.g., show a spreadsheet for the budget and demonstrate the technology). It is only when there is a need for also sharing an embodied and affective understanding within the team, that the sensitizing scenarios become a suitable form. Concepts from the social sciences, e.g. sociology, anthropology and museum studies, are suitable targets as they relate to phenomena that manifest in everyday life, in social interaction between people, and in culturally conditioned practices. It makes sense to re-enact these as a way to get an embodied and affective relation to the concepts, to complement an otherwise primarily analytical perspective. Finally, scenarios should thus not be seen as a stand-alone method, but must be combined with other methods to make for a useful design exercise total.

Facilitation

The person facilitating should preferably themselves have played the scenario at least once. This allows them to understand the structure of the scenario and be better aware of how to pace the different phases (when to slow down and speed up its execution). Experienced facilitators are typically able to pick up a short scenario and run it without having played it themselves. For the inexperienced facilitator, we include some basic tips below.

- Read through the whole scenario carefully before staging it. Make sure you have done all preparations before the players arrive.
- Distribute roles randomly if you do not have strong reasons to do anything else. For example, it's better to not cast players in roles that are too similar to their project function (in particular, avoid casting managers as managers) and never cast players based on age, race, or gender.
- If you can, use simple costuming to help players develop their characters. A useful method is to bring a collection of random accessories and pieces of clothing, and let players pick and choose what they want to wear to express their character. If this is not possible, you can just give them a moment to adjust clothing, hair, etc., to slightly change their appearance.
- Minimize space. Forcing players to be physically close to each other increases the chances that they will interact with most of the other characters. Conversely, in situations where you want the characters to *not* interact, distribute them physically (e.g., in separate corners).
- Develop a sense for pacing. For example, preparations have a tendency to drag out in time and the facilitator must keep the tempo up to not run out of play-time. In scenes where players have nothing else to do than talk to each other, five minutes is already a very long time. Cut such scenes short, unless they are *meant* to be awkward. On the other hand, for scenes where the characters are working towards a joint goal, fifteen minutes will often feel rushed.

The Quest for Authenticity

There is a risk that your participants will not find the scenario sufficiently *authentic*. We have heard this concern raised about everything from the setting (e.g., if the museum in the scenario is fictional), the time frame allowed for scenes (e.g., when a process should take days, not minutes), the characters, and the first-person experience of stepping into a role. While the concern is valid, the sensitizing scenarios are not meant to be, and in fact cannot be, accurate simulations of authentic situations. They serve as reference points for discussion and reflection and to that purpose, a more abstract scenario works as well (or better) than an accurate simulation.

It is better that authenticity is provided *by the participants*. One way this happens is during the debrief, as the participants re-connect to their own design task and context. It is also possible to heighten the connection by adapting a scenario to its specific context of a design project.

One more way to heighten authenticity is to let players play characters that lie close to themselves. Players can, for example, tap into their professional skills to create a higher level of authenticity. This approach should however be used with care, since it can lead to players experiencing performance anxiety around doing a 'good job'. The characters created for *The Object* open up for both: While players are free to play on their professional capabilities, they are at the same time asked to adopt *attitudes* that are not their own. This will for example let the museum pedagogue play a museum pedagogue, but with a different attitude towards the collections than their own.

Finally, one should *never* expect that role-playing a character means knowing what it is to be that person!! Role-playing comes with the risk of 'identity tourism' (Nakamura, 1995), if participants enact their characters in ways that are racially or otherwise stereotyped. Since sensitizing scenarios are short and abstracted, some stereotyping will always come into play. Players must always be cautioned about these issues and encouraged to play them with respect.

Finding and Adapting Scenarios

When running sensitizing scenarios, it is important to use a scenario that makes sense for the design project at hand. Many times, it also makes sense to do some adaptations to the scenario to fit with the particular museum or project design goals. For example, *The Object* was staged as part of a design workshop in the Museum of Yugoslavia.[10] In this setting it was useful to pick a typical object from their collection as the object of the scenario.

Of the scenarios developed within Gift, *The Object* and three other scenarios have been made available from the GIFT website.

- *My Museum* is designed to be played in an authentic museum exhibition, with the purpose of understanding group interactions and the target audience better. It can be tailored to become an evaluation tool, by staging it in a new exhibition or using a prototype hybrid experience.
- *The Gift* illustrates the complexities of human gift-giving practices. The scenario plays out in a family gathering, and some of the gifts illustrate how museums can become involved in gifting practices.
- *The Race* illustrates theories about player motivations for social play in games, and is useful if a project intends to use games or encourage play in the museum context. This scenario includes a board game as well as role-play, and is the most abstract of the four scenarios.

In addition, there exist a plethora of scenarios that have been developed for educational and artistic purposes. Many are freely available. These can function as sensitizing scenarios, but are typically much longer than those developed within GIFT, and may pose requirements on the space for play or require props. For practical purposes, they might need to be shortened or simplified. They also very seldom include a structured debrief, at least not for the purposes of running them for a design project. This means that the facilitator will have to develop their own debrief method, a task that should not be taken lightly. Finally, the facilitator must take the number of players into account – some scenarios are playable only with certain numbers of players.

Sources for freely downloadable scenarios include the following websites.

- Stockholm Scenario Festival (https://scenariofestival.se/) features short scenarios that are light-weight on rules and often centre on serious themes. Most are available in English from the website and downloadable for free.
- RPGNet (https://wiki.rpg.net/index.php/LARP_Scenarios) features a collection of small role-play scenarios that are freely downloadable. This collection has a stronger focus on scenarios that are primarily meant to entertain and some are quite rule-heavy.

10 Waern and others, 'Sensitizing Scenarios'.

- Interactive dramas (http://www.interactivedramas.info/scenario.htm) hosts a collection of what they call 'interactive dramas' as a way to connect to theatre. Most are available for free.
- Black & Green Games (http://www.blackgreengames.com/) is an independent game publishing imprint. The site focusses on role-playing games available for purchase but also have some that are freely downloadable. The site also has feature articles.

Some specific example scenarios that may be useful for certain museum design projects include:

Public Memory.[11] A scenario about what happens when your views about history are challenged. It seeks to bring the subtleties of institutional racism to light, and wrestles with the ideas of who gets to decide how history remembers a person, and what to do when that person is complicated – neither good nor evil.

Here Is My Power Button[12] is an intimate near-future scenario about the interaction between humans and future, near-intelligent AI building their personhood through interacting with humans.

Sign.[13] A card game about how language can develop as collectively constructed.

Artsedge.[14] A collection of scenarios teaching arts and theatre literacy and technique.

Feminism.[15] A collection of very short scenarios, nano-games, grappling with contemporary feminist issues.

Designing Your Own Scenarios

Designing role-playing scenarios is at the same time very easy and enormously difficult. The reason for both is the same: It is to a large extent the

11 https://learnlarp.com/ (Accessed December 2020).

12 https://batwater.itch.io/here-is-my-power-button (Accessed December 2020).

13 https://thornygames.com/pages/sign (Accessed December 2020).

14 https://artsedge.kennedy-center.org/multimedia/Interactives/artsedge-games/170224-artsedge-games.aspx#ae-games-rpg (Accessed December 2020).

15 https://feministnanogames.wordpress.com/ (Accessed December 2020).

players that create the scenario by acting it out. Scenarios require minimal resources to design and to stage; many consist in their entirety of a printed instruction for the facilitator and some printouts. At the same time, the same scenario can play out very differently depending on how the participants choose to engage with it, and very small differences in design can have huge effects on how players choose to engage. The best analogy is probably that of a chaotic system (Montola, 2004), in how any detail can create spin effects that affect their execution. For these reasons, it is better to use and adapt existing scenarios that have been tested and tried, than to design entirely new ones.

If you nevertheless decide to design a new scenario, there are some design strategies that may help. Firstly, since sensitizing scenarios have an educational purpose, designers will typically try to exert some level of control over this potential chaos. One way is to divide the scenario into multiple short and scripted scenes, as illustrated by *The Object*. The presence of a facilitator is also a stabilizing factor, as the facilitator can cut into play and do adjustments on the fly.

An important consideration relates to the characters. For sensitizing scenarios, we recommend using fictional characters who come with their own attitudes and goals, that are different from those of the player. Using fictional characters creates an alibi for acting in unusual ways and relieves the participants of having to perform well in a given task. However, characters can be short and rudimentary described, letting players flesh them out during pre-play exercises included in the scenario. It is however important to be explicit about those details that are important for how the scenario is intended to play out. For example, if it is important that an object is destroyed at the end of a scene, this must be explicitly stated in the instructions to the facilitator.

Finally, for practical purposes, sensitizing scenarios must be kept short (maximum 1.5 hour and ideally no longer than 45 minutes) and they must support a varying number of players. Else, most projects will find it very difficult to adopt them.

Resident Theorist

Recruiting a *Resident Theorist* (further RT) to be part of a design team is another way to foreground the development of a common theoretical grounding. This is a functional role within a project team, a person who keeps track of how the team develops their terminology and making sure that team members share a somewhat common understanding of concepts that are critical for the project focus and direction. The RT is tasked with provoking the project team with relevant concepts and theories deemed relevant by the team itself, to heighten their critical awareness of both misunderstanding and critical gaps in their reasoning.

The role can be scaled to the project it is part of. In smaller groups with limited resources, it can be filled by a single person whereas in larger groups it can be distributed among a group with various members focusing on particular goals. It needs to be filled by a researcher or another person comfortable in interdisciplinary theoretical concepts and theories. The main requirement for the role is a thorough knowledge of concepts relevant within the project. The role can be taken by a person dedicated to this purpose, or by a person who also has other roles in the project. However, in order to be able to maintain critical distance to the issues that arise, it is advisable that the person does not have direct decision-making power within the design process.

Within GIFT one of the authors (Rajkowska) was tasked with acting as RT, in addition to other tasks within the project. The role was adopted from the start of the project and lasted throughout the process, with focus on the project-internal design processes. The discussion below is based on the experiences from this process.

The RT intervention is best described as a form of action research, and as such is closely related to the methods and processes discussed in Chapter 6. As in all action research, acting as RT requires a constant circle of personal experience, action, and reflection on the result of the action.[16] Action research will always focus on instigating the change process, grounded in the personal experiences of project members and their task, including the resident theorist. As every project is different and encounters different issues, the RT must always adapt their intervention to the situation at hand and carefully reflect on the change processes they help catalyse. It is also important to approach the change process as one that is owned by project participants, and not one in which the Resident Theorist themselves have a strong stake. Reflection allows the Resident Theorist to learn from their own practices, improve between cycles, and be transparent in accounting for their interventions.

16 Adelman, 'Kurt Lewin and the Origins of Action Research'; Lewin, 'Action Research and Minority Problems'.

Reasons for interventions

Within GIFT, we saw two main reasons for when the resident theorist decided to intervene. Interventions were spurred by either a *lack of common vocabulary* or a need to *surface tensions.*

Providing Vocabulary

This situation arises primarily when there are terminology differences between different domains of work and research. The role of the RT is in this case to recognize when a term is being used that only part of the team may be fully versed with, as well as recognize when different team members may be using the same words with different meanings and/or connotations. Within GIFT a recurring example (in several subprojects) was the concept of *appropriation*, a word that had been used to denote a desirable effect in the project plan. For interaction designers, the concept of appropriation is a largely positive term which relates to the ability of a user to adapt the content and use of a design to fit their own perspective or purpose, or to create something new out of a given material.[17]

Among museum theorists and practitioners, appropriation is however typically discussed with a negative connotation, as the insensitive inclusion of (objects from) other cultures into western-centric exhibits.[18] In the museum context, appropriation thus tends to be a delicate topic and considered undesirable. The different connotations to the term, as positive or negative, created tension within discussions during an initial workshop in the project. The RT was able to identify the issue due to being versed in both discourses. The RT then asked for clarification of the concept and the way it is being currently applied. This prompted a shift in the ongoing discussion, towards a desire to agree on common meanings for contested words. The designers and the museum representatives then went on to clarify what they were talking about, when talking about appropriation. In the end a new meaning of the word appropriation was negotiated for use within the project, agreeing that it would be understood as adding a new layer of meaning to existing objects:

> Developer: Basically, you see this ashtray that was bought for 2.99 dollars on a flea market and then William Gibson wrote a story, and all stories

17 Dix, 'Designing for Appropriation', 11.

18 Pearce, *Museums and the Appropriation of Culture.*

> are just one page, so, one page. It's a plastic horse, an airplane, trash, basically trash. But when writers made narratives those objects became rich objects and that completely changed the character of the object. That's appropriation of the object. So they appropriated the object with their narratives.

A key factor in being able to maintain a constructive discussion in this case was that the RT could recognize the origins of concepts and terms. Also, being open towards terms having different meanings and connotations in different knowledge domains was instrumental in allowing the team to jointly uncover their own, relevant, meaning for the design context at hand.

Bringing Tensions to Surface

Another reason for the resident theorist to intervene was related to bringing tensions to the surface, that had previously been left tacit. The developers and researchers had been strongly inspired by themes of playfulness and games within the space of museums, while the museums raised questions related to what the overarching educational purpose of those themes would be. This led to a standstill in the discussion, where each side wanted to emphasise their point of view, on how to put the priorities in the design process. The discussion of educational values versus fun can be discussed through the lens of the struggles that new museology went through when it first emerged almost 30 years ago, as problematized by Hooper-Greenhill.[19]

Recognizing the issue encouraged the RT to intervene. In order to highlight the tension, the RT rearticulated the arguments from both parties to clarify if they were correctly understood, and then rephrased them in reference to museum studies literature. In doing so, the arguments became rephrased using a vocabulary characteristic of new museology studies. The RT further emphasised that such conflicting goals have a long history in museum research, and that they have been managed by museum organizations in similar situations before. In that way, the RT lifted the discussion to a more general level and made it apparent that this was not just a problem within this particular project and context.

Presenting the tension as one commonly encountered allowed for an easier exchange of concerns. The intervention reframed the problem, and allowed the participants to take a step back, find composure and discuss

19 Hooper-Greenhill, 'Audiences: A Curatorial Dilemma'.

the issues in a more structured and transparent manner. While the issue was not resolved, these discussions carried weight in subsequent workshops. The tension continued to be openly addressed during the following testing and feedback sessions, where educational value and fun factors were both considered to be relevant aspects of success.

Again, we can see how the resolution was dependent on the RTs ability to recognize the tension and raise the discussion to a more generic level, allowing the parties to take a step back and acknowledge the value of both positions.

Strategies and Tactics

The RT in GIFT developed strategies for both informal and pre-planned interventions. The informal ones basically consisted of the RT speaking up when participating in a meeting or workshop. As for pre-planned interventions, the RT prepared a presentation on one occasion, and contributed to the structure and content for a design workshop on another.

It is important here to mention that inclusion of the RT was something supported by the team. The theories that were used were selected during a theory workshop and then left to the RT as resources that connect to the value systems of other participants. In that sense, the RT needs to be comfortable navigating among the different theories but the role also lifts the already existing expertise of the team and makes sure that one of the project team members is explicitly focused on this critical reflection and theory inclusion when others must focus on constantly shifting tasks.

Being an RT is visibly about theory but it is primarily about being a resource to others and contributing to the underlying process with reflection while never stepping in to impose particular choices. As mentioned previously, in this case the role was filled by a PhD student. Due to how PhD programs are laid out in Scandinavian countries, she had the time to familiarize herself with the theories before starting to participate in the various activities. That said, being still a student is the lowest step of an academic career and it can be difficult to provide a critical reflection to people who have decades of experience. Resident Theorist is a member of the team and the initiative to include theory and reflect on practice must be a team decision for it to work. As with any action research process, this is not straightforward and each project will have to make its own adjustments to the method but we believe this is a viable avenue towards disseminating sensitizing concepts.

Other Ways to Sensitize Designers

This chapter has presented two specific methods for sensitizing designers towards complex and nuanced theories, with the goal of keeping the design process focussed on key goals and potential issues. The presented methods have been tested and tried within GIFT and associated museum development projects with reasonable success.

Given that both methods can be considered too resource-consuming, we want to conclude this chapter by presenting options for how similar awareness can be accomplished through simpler, but also perhaps less effective, means.

Keeping and maintaining a glossary

A glossary helps a team develop a joint agreement on the meaning of core terminology, as well as on what terminology is important. The method combines well with the Resident Theorist method as the glossary can be maintained by the RT. However, it risks alienating those members of the team that use other terms, or the same terms with slightly different meaning, in their own skilled practice.

Mood boards

Mood boards are collections of images, made accessible to a project team as a joint reference point. They are frequently used in graphics design to set the graphical theme of a product. It is also possible to use mood boards to create a wordless shared reference to social contexts and experiences, and they can also serve as a basis for discussion. Mood boards may be particularly useful in the context of Hybrid Museum Experiences, as they are able to merge and bridge physical and digital. Just as a glossary, a mood board can be maintained by a RT, but it is more common for mood boards to be developed collectively by an entire project team.

Participatory design, joint observation and analysis

Many museum projects are directed towards specific audiences, and include the target audience in the design process. This can be done through ethnographic observation, or by directly inviting members of the target audience to participate in the design process. While participatory design methods are at least as resource-consuming as those discussed in this chapter, they can be adapted to achieve two things at once: They can allow the team to understand their audiences better as well as sensitize the team towards audience-related concepts and theories. In order to tap into this opportunity, it is important that the entire team gets a chance to meet and interact with the target audience, and collectively work towards developing key sensitizing concepts.

Bibliography

Clem Adelman, 'Kurt Lewin and the Origins of Action Research', *Educational Action Research*, 1.1 (1993), 7–24 <https://doi.org/10.1080/0965079930010102>

Adam Blatner, *Interactive and Improvisational Drama: Varieties of Applied Theatre and Performance* (iUniverse, 2007)

Herbert Blumer, 'What is Wrong with Social Theory?', *American Sociological Review*, 19.1 (1954), 3–10 <https://doi.org/10.2307/2088165>

Alan Dix, 'Designing for Appropriation', in *Proceedings of the 21st British HCI Group Annual Conference on People and Computers: HCI... But Not as We Know It* (British Computer Society, 2007), II, 27–30

Thomas Duus Henriksen, 'Learning By Fiction', in *When Larp Grows Up: Theory and Methods in Larp*, ed. by Morten Gade, Line Thorup, and Mikkel Sander (Frederiksberg: Projektgruppen KP03, 2003), pp. 110–115 <https://nordiclarp.org/w/images/c/c2/2003-As.Larp.Grows.Up.pdf> [accessed 4 June 2021]

Eilean Hooper-Greenhill, 'Audiences: A Curatorial Dilemma', in *The Educational Role of the Museum*, ed. by Eilean Hooper-Greenhill (London & New York: Psychology Press, 1999), pp. 255–268

Kurt Lewin, 'Action Research and Minority Problems', *Journal of Social Issues*, 2.4 (1946), 34–46 <https://doi.org/10.1111/j.1540-4560.1946.tb02295.x>

Susan M. Pearce, *Museums and the Appropriation of Culture* (The Athlone Press, 1994)

Susan Leigh Star, 'This is Not a Boundary Object: Reflections on the Origin of a Concept', *Science, Technology, & Human Values*, 35.5 (2010), 601–617 <https://doi.org/10.1177/0162243910377624>

Susan Leigh Star, and James R. Griesemer, 'Institutional Ecology, 'Translations' and Boundary Objects: Amateurs and Professionals in Berkeley's Museum of Vertebrate Zoology, 1907-39', *Social Studies of Science*, 19.3 (1989), 387–420 <https://doi.org/10.1177/030631289019003001>

Annika Waern, Paulina Rajkowska, Karin B. Johansson, Jon Bac, Jocelyn Spence, and Anders Sundnes Løvlie, 'Sensitizing Scenarios: Sensitizing Designer Teams to Theory', in *Proceedings of the 2020 CHI Conference on Human Factors in Computing Systems*, CHI '20 (New York, NY, USA: Association for Computing Machinery, 2020), pp. 1–13 <https://doi.org/10.1145/3313831.3376620>

About the Authors

Paulina Rajkowska is a PhD Candidate at the Department of Informatics and Media at Uppsala University. She comes from a mixed background of sociology, media studies and HCI. Her prior work includes writing on social phenomena in video games as well as application of critical theory to design practices. In her ongoing research she is looking at reflexive practice, playfulness and how both of those can be applied in projects, to improve the quality of their outcomes.

Annika Waern is a 'research through design' academic who has been researching technology-supported physical play and games for about fifteen years. Currently, she is conducting research in the areas of hybrid play in museums, children's play in outdoor settings, and circus training to foster proprioceptic skills. Waern has authored over 100 academic articles and book chapters, and been editor of 2 books and several journal special issues. She is a DiGRA (Digital Games Research Association) distinguished scholar and a HEVGA (Higher Education Video Game Alliance) fellow.

8. Ideation Tools for Experience Design

Anne Rørbæk Olesen, Christian Hviid Mortensen, Anders Sundnes Løvlie

Abstract

This chapter argues for the importance of ideation in design processes. We present three paper-based tools for ideation to be used at a meeting or a workshop tailored specifically to the museum sector and developed in collaboration with museum professionals. Each tool targets a different phase of the design process. There is a tool for generating ideas, strengthening ideas and testing ideas. The chapter accounts for the what, when, how and why of these ideation tools. Finally, we discuss how the tools can be used in combination and how they can encourage reflection.

Keywords: Ideation; Experience design; Design processes; Museum technology; Action research; Theory of change.

As was argued in Chapter 1, the perspective we offer in this book gives primacy to experiences over technologies. We are not proposing technology for the sake of technology, but rather as a potential route towards fulfilling a purpose that is relevant and meaningful in a specific museum for particular visitors. Therefore, design should not be driven or dominated by technology but by what it seeks to accomplish – by the *idea*. The processes of generating, developing and communicating ideas, then, becomes crucial through what has been called *ideation*.[1] While ideation is always an important element in design, we find it particularly important to prioritize in designing hybrid museum experiences.[2]

1 Jonson, 'Design Ideation: The Conceptual Sketch in the Digital Age'.

2 Laamanen and Seitamaa-Hakkarainen, 'Interview Study of Professional Designers' Ideation Approaches'.

Waern, A. and A. Sundnes Løvlie (eds.), *Hybrid Museum Experiences: Theory and Design*. Taylor & Francis Group, 2022
DOI 10.5117/9789463726443_CH08

The museum domain offers many challenges for hybrid design. The use of technology often requires external collaboration, e.g. with designers and developers that are not part of the museum staff. The work of crafting visitor experiences through collaboration across museum professionals from a variety of departments and disciplinary backgrounds – curators, educators, communicators, etc. – is further complicated by collaboration with external parties. Much research has explored the challenges that arise in such collaborations.[3] In particular, facilitation of collaboration between professionals from different backgrounds raises the need for robust and flexible ideation tools.

In this chapter, we showcase three ideation tools developed with and for museums as a way to help prioritize ideation in early phases of design work. They are paper based tools that are intended to be used at a meeting or workshop to support collaborative generation of new ideas or the development of existing ideas.

The Importance of Ideation

As noted by Löwgren and Stolterman 'every design starts out as an idea'.[4] Following Jonson, an idea can be understood as 'a basic element of thought that can be either visual, concrete or abstract'.[5] But where do ideas come from? And how do they become 'designs'? While there are no simple answers to these broad questions, one thing is certain: Ideas need work. As suggested by Löwgren and Stolterman, ideas should be transformed, externalized and made visible in order to be collaboratively criticized, developed, expanded, revised, and often discarded.[6] These kinds of ideational tasks are typically attributed to or explored in the early phases of design, focusing on the generation and emergence of ideas.[7] However,

3 Knudsen and Olesen, 'Complexities of collaborating'; Olesen, 'Mapping Innovation Processes: Visual Techniques for Opening and Presenting the Black Box of Service Innovation Processes'; Olesen, Holdgaard, and Laursen, 'Challenges of Practicing Digital Imaginaires in Collaborative Museum Design'; Parry, *Recoding the Museum*; Peacock, 'Making Ways for Change'.

4 Löwgren and Stolterman, *Thoughtful Interaction Design: A Design Perspective on Information Technology.*

5 Jonson, 'Design Ideation: The Conceptual Sketch in the Digital Age', p. 13.

6 Löwgren and Stolterman, *Thoughtful Interaction Design: A Design Perspective on Information Technology*, p. 51.

7 Dorta, Pérez, and Lesage, 'The Ideation Gap'; Halskov and Dalsgaard, 'Inspiration Card Workshops'; Laamanen and Seitamaa-Hakkarainen, 'Interview Study of Professional Designers'

ideation very often improves through physical manifestation, which is why prototyping and early evaluations (see Chapter 10) are also recognized as important ways to work with, test and iterate ideas on their route towards becoming designs.[8]

Ideation is clearly important. But, actually, we did not realise *how important* it was in the context of museums when we initiated the GIFT Action Research Module – a 1.5 year long process with ten museums from Europe and USA (see Chapter 6), building on Culture24's *Let's Get Real* action research methodology.[9] The tools presented in this chapter were used in and grew out of this process. They were inspired by previous research on designing digital technologies at museums that showed a tendency towards tech-driven development.[10] We therefore set out to prioritize initial phases of design, valuing contextual and collaborative activities. The result of this process was however much more radical than we originally imagined, in that many of the museums questioned why the process needed to be framed through the perspective of technology. Three practices were highlighted as important: 1) to start idea generation with purpose and people in mind and not technology, 2) to collaborate on ideas, and 3) to test ideas. The tools presented in this chapter seek to support these ideational practices in a museum context.

Three Tools for Ideation

The three tools that we showcase are called the *VisitorBox Ideation Cards*, the *ASAP Map* and the *Experiment Planner*. They are all intended to support collaborative work but at different stages of ideation: The *VisitorBox Ideation Cards* (Figure 8.1) focus on generating ideas, the *ASAP Map* focuses on strengthening ideas and the *Experiment Planner* focuses on testing ideas.

Ideation Approaches'; Wetzel, Rodden, and Benford, 'Developing Ideation Cards for Mixed Reality Game Design'.

8 Sanders and Stappers, 'Probes, Toolkits and Prototypes'.

9 See https://weareculture24.org.uk/lets-get-real/.

10 Olesen, 'For the Sake of Technology? The Role of Technology Views in Funding and Designing Digital Museum Communication'; Olesen, Holdgaard, and Laursen, 'Challenges of Practicing Digital Imaginaires in Collaborative Museum Design'.

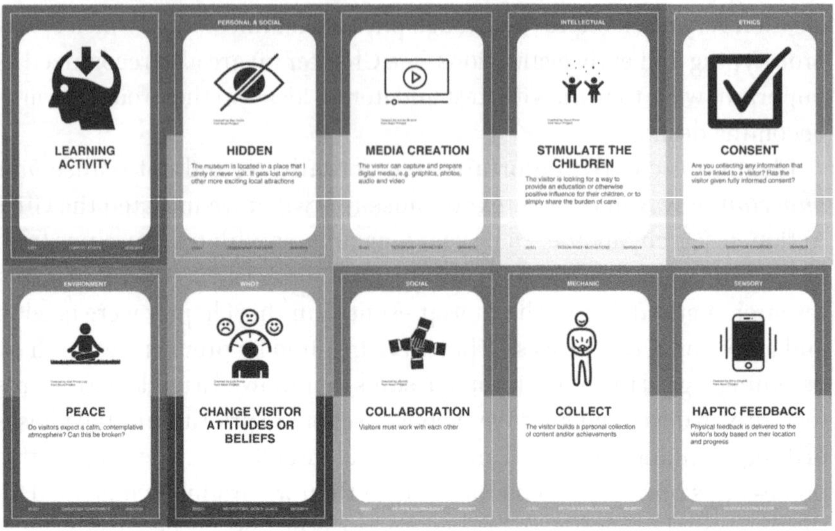

Figure 8.1 VisitorBox Ideation Cards – a small selection. You can download the entire deck as a printable pdf at https://visitorbox.org/

What? VisitorBox Ideation Cards is a card game that helps you come up with innovative and thoughtful ideas for digital experiences.

When? Play the cards in the early phases of design as a fun and inspiring method for generating new ideas for digital experiences.

How? Print the cards and bring them to a meeting with relevant collaborators.

Why? Museums that use the cards generate innovative ideas and are able to demonstrate exactly how they determined an idea to be 'good' or 'bad.'

The VisitorBox Ideation Cards: Generate Ideas

The purpose of ideation cards is to support collaborative design in a playful way. Many such ideation card decks have been developed, for different purposes and contexts.[11] They typically encode important design knowledge in a domain and suggest a range of design options of particular relevance.

The *VisitorBox Ideation Cards* were developed specifically for cultural heritage institutions by Ben Bedwell and colleagues at the University of Nottingham, and focus on the use of various technologies to support visitor experiences. The card deck incorporates knowledge derived from a broad range of design projects in cultural heritage institutions. The cards are designed to help identify a number of issues that are relevant in ideation

11 Wetzel, Rodden, and Benford, 'Developing Ideation Cards for Mixed Reality Game Design'.

– some of which have to do with technologies for hybrid experiences, but also a variety of other relevant concerns such as understanding the audiences, the institution's main assets, goals, barriers, etc. You start the game by thinking about the purpose of designing an experience and the people you want to reach, through discussing context and institutional goals. Next, you create a design brief, leading on towards generating and storyboarding design ideas. As a final stage, you take a critical look at your idea by considering a number of 'disruptive' cards, in order to determine whether the idea is strong enough to pursue further.

Using the cards became an important part of the work done by museum professionals in the GIFT Action Research. In evaluation, one participant said that the cards were 'a great exercise for thinking concretely about tech'. Some of the participants implemented the method in their home institutions afterwards. As a museum professional stated later: 'I really found the cards useful. I have used them loads since then and they really help you design a design brief and then think about what you can do and how you can do it.'

The ASAP Map: Strengthen Ideas

What? The *ASAP Map* helps you facilitate a discussion about an idea in order to develop shared understanding and build on what you already know.

When? Use the map as soon as possible when you have an idea for a digital experience.

How? Print the map and use it to facilitate a discussion at a meeting.

Why? Use the map to strengthen your idea, make on-going collaboration easier and be able to better explain your decisions.

The *ASAP Map* is inspired by the idea of making maps for design reflection, as a reflection-in-practice intervention that supports people in reflecting collaboratively on the context in which they design and on which way they are heading.[12] You use the map when you already have a design idea that you would like to develop further with relevant collaborators, focusing in particular on *the purpose* behind the idea. First, you discuss the purpose and then talk about it through questions split into four categories: Awareness, Solutions, Alliances and Plans (ASAP). The name of the method, *ASAP Map*, both refers to the four categories, but also playfully encourages people to use it ASAP (As Soon As Possible) when having some kind of first idea.

12 Dalsgaard, Halskov, and Nielsen, 'Maps for Design Reflection'; Schön, *The Reflective Practitioner: How Professionals Think in Action.*

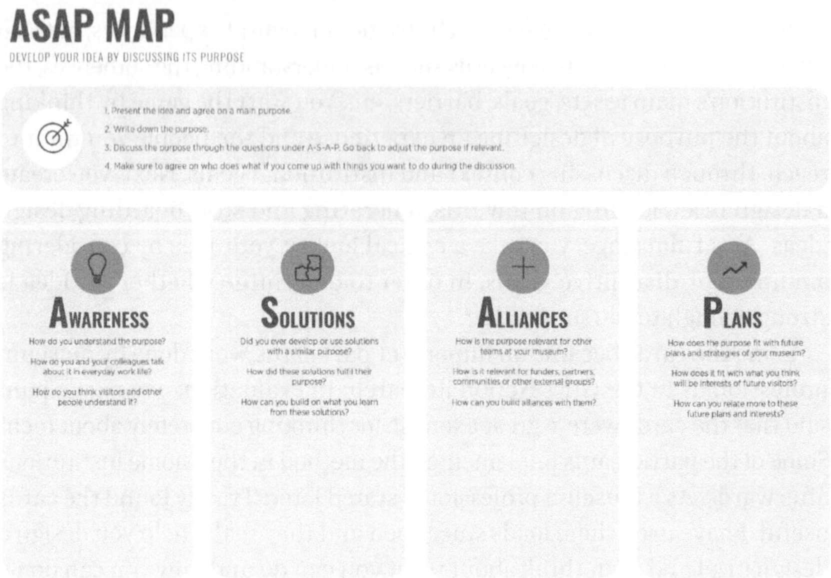

Figure 8.2 The ASAP Map. You can download the map as a printable pdf from http:gifting.digital.

The *ASAP Map* was developed in the GIFT Action Research activity. Early in the process, the ten museum professionals tested an initial version of the map in their home institutions. Through this test and as a result of two follow-up workshop sessions with the group, the map was iteratively developed into the final version. From the tests with the first version we could see that the map held a potential to support both micro level discussions on new perspectives, possibilities or barriers and more macro level discussions on strategies and strategic awareness. In the workshop discussions, the museum professionals highlighted the map as a way to 'dig in and give flesh and bone to some fancy stuff' and as a good frame for discussions: 'If somebody comes up with a digital idea, you can use the map to anchor it.' Also, they saw it as having a potential for supporting early collaboration, as one of the museum professionals argued: 'Without all stakeholders involved from the very start, you're building in a potential problem further down the line.' A more detailed presentation of the ASAP map and the design process can be found in a recent study by Olesen, Holdgaard and Løvlie.[13]

13 Olesen, Holdgaard, and Løvlie, 'Co-Designing a Co-Design Tool to Strengthen Ideation in Digital Experience Design at Museums'.

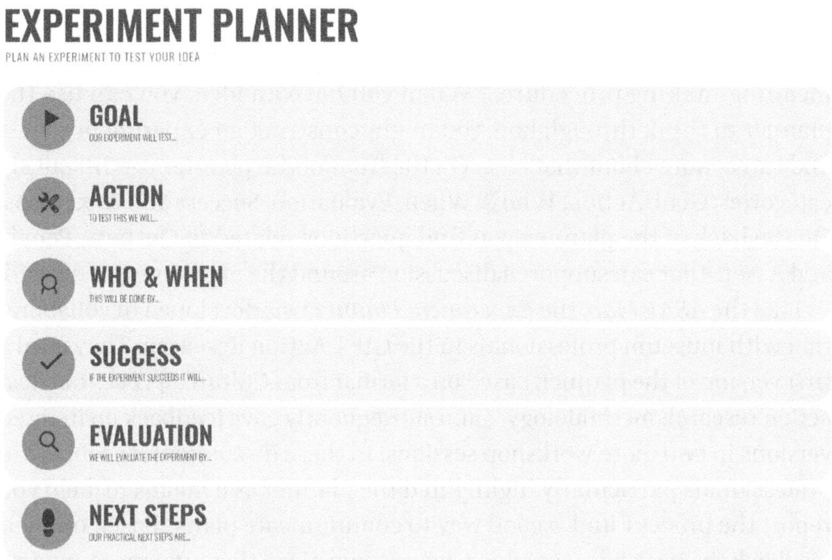

Figure 8.3 The current version of the Experiment Planner. You can download the planner as a printable pdf at http://gifting.digital.

The Experiment Planner: Test Ideas

What? The *Experiment Planner* helps you plan an experiment to find out if your idea makes sense and how you can develop it.

When? Use the *Experiment Planner* when you have an idea that you would like to test or know more about.

How? Print the planner and bring it to a meeting with relevant collaborators.

Why? Use the planner to save time and money, make more useful solutions and be able to better explain your decisions.

The *Experiment Planner* encourages experimental practices. The method is inspired by principles from Theory of Change,[14] where the idea is to explicate one's theory about how and why an initiative might change an existing situation. Doing so helps to articulate what might be done in order to enable the change, and later on evaluate whether the desired change was

14 Connell and Kubisch, 'Applying a Theory of Change Approach to the Evaluation of Comprehensive Community Initiatives: Progress, Prospects and Problems'; Weiss, 'Nothing as Practical as Good Theory: Exploring Theory-based Evaluation for Comprehensive Community Initiatives for Children and Families'.

realised or not. With the Experiment Planner, the focus is not on changing existing situations but on experimentation as a knowledge-generating or meaning-making procedure.[15] When you have an idea, you can use the planner to think through how you might construct an experiment to test and learn more about that idea. On the front of the planner you fill out six categories: Goal, Action, Who & When, Evaluation, Success and Next Steps. On the back of the planner, you find questions related to Outputs, People and Assets that can support a discussion around the experiment if needed.

Like the *ASAP Map*, the *Experiment Planner* was developed in collaboration with museum professionals in the GIFT Action Research. They used a first version of the planner, based on a format from Culture24's *Let's Get Real* action research methodology,[16] and subsequently gave feedback on iterated versions in two more workshop sessions. In these discussions, the museum professionals particularly highlighted the planner as a means to 'help you report the process' and 'a good way to communicate plans'. Based on their feedback the backside was added, having questions that offer more support in filling out the planner. This was deemed useful later in the process, as one of the museum professionals expressed: 'I really like the *Experiment Planner*. I think it works really well. Having the questions on the back, those prompt questions, I think that is really helpful.'

Combination and Usefulness of the Tools

In this chapter, we have showcased three paper based ideation tools developed with and for museums with the purpose to help prioritize ideation in early phases of design work. As a concluding reflection, we would like to ponder a bit on the possible combination of these tools as well as their usefulness.

It is important to say that each project should find its own way of combining them. While the three tools surely can be used one after the other in the presented sequence (from *VisitorBox Ideation Cards* via the *ASAP Map* to *Experiment Planner*), they could also be used separately or combined with other approaches. Acknowledging the differences across museum institutions and practices, the tools should be used in whatever way they might make sense in a specific context. In our test of the first version of the *ASAP Map* we found that museum professionals used the map quite differently in their home institutions: There were significant variations in

15 Macdonald and Basu, *Exhibition Experiments*.

16 See https://weareculture24.org.uk/lets-get-real/.

terms of who participated in using the map, the setting of use and how the map was used. Overall, this observation illustrates the flexibility of the map.[17]

This leads us to reflect on the usefulness of the ideation tools. Following Löwgren and Stolterman,[18] we acknowledge that 'normative approaches are not enough. In order to handle the complexity of interaction design, there is a need for a reflective mind – what we would label a thoughtful designer.' The tools cannot, and are not intended to, substitute a reflective mind. They are meant as a way to encourage and support reflection – not only for the singular mind but as a collaborative endeavour. However, this is not an easy task. Especially not since museums often include a wide range of professions and different levels of digital literacy. As a museum professional in the GIFT Action Research explained: 'I would see digital as being more experimental, thinking about design practices, being more agile, taking more risks, whereas the people in the organization see it very much as technology led.'

In a similar vein, another museum professional highlighted the need for deeper organizational changes in order for the ideation tools to work: 'To actually be accepted as a way of doing things, you would need to have the right capacity.' However, he did find them 'a useful first' that 'does create space' for reflection. Effective organizational change starts with one person; one reflective mind who is willing to take up the challenge to include others in the reflections. Taking on that role is a challenging task, as a museum professional expressed:

> All these big companies are constantly selling you products from the basis that 'this product solves that problem'. And so many people are acculturated to assuming that's the way digital works. So, if you're a person on the move who thinks 'well, we could do this but we need to do x, y and z', you're always in that disadvantage where you present the most complicated set of arguments.

While this is not an easy role to take on, it might be necessary. Following the arguments advanced in this chapter, there is a need to carve out space for ideation when designing hybrid museum experiences. The three ideation tools presented might support or inspire ways to achieve this goal, hopefully resulting in museum experiences that are driven by ideas over technologies.

17 Olesen, Holdgaard, and Løvlie, 'Co-Designing a Co-Design Tool to Strengthen Ideation in Digital Experience Design at Museums'.
18 Löwgren and Stolterman, *Thoughtful Interaction Design: A Design Perspective on Information Technology*, p. 2.

Bibliography

James P. Connell, and Anne C. Kubisch, 'Applying a Theory of Change Approach to the Evaluation of Comprehensive Community Initiatives: Progress, Prospects and Problems', in *New Approaches to Evaluating Community Initiatives: Theory, Measurement and Analysis*, ed. by Karen Fullbright-Anderson, Anne C. Kubisch, and James P. Connell, 2 vols. (Washington, DC: The Aspen Institute, 1998), II, 15–44

Peter Dalsgaard, Kim Halskov, and Rune Nielsen, 'Maps for Design Reflection', *Artifact*, 2.3–4 (2008), 176–189 <https://doi.org/10.1080/17493460802526412>

Tomás Dorta, Edgar Pérez, and Annemarie Lesage, 'The Ideation Gap: Hybrid Tools, Design Flow and Practice', *Design Studies*, 29.2 (2008), 121–141 <https://doi.org/10.1016/j.destud.2007.12.004>

Kim Halskov, and Peter Dalsgaard, 'Inspiration Card Workshops', in *Proceedings of the 6th Conference on Designing Interactive Systems*, Proc. of Designing Interactive Systems (New York, NY, USA: ACM, 2006), pp. 2–11 <https://doi.org/10.1145/1142405.1142409>

Ben Jonson, 'Design Ideation: the Conceptual Sketch in the Digital Age', *Design studies*, 26.6 (2005), 613–624

Line Vestergaard Knudsen, and Anne Rørbæk Olesen, 'Complexities of Collaborating: Understanding and Managing Differences in Collaborative Design of Museum Communication', in *The Routledge Handbook of Museums, Media and Communication*, ed. by Kirsten Drotner, Vince Dziekan, Ross Parry, and Kim Christian Schrøder (London and New York: Routledge, 2019), pp. 205–218 <https://gifting.digital/wp-content/uploads/2020/01/Preprint_complexities-of-collaborating-1.pdf> [accessed 29 October 2019]

Tarja-Kaarina Laamanen, and Pirita Seitamaa-Hakkarainen, 'Interview Study of Professional Designers' Ideation Approaches', *The Design Journal*, 17.2 (2014), 194–217 <https://doi.org/10.2752/175630614X13915240575988>

Jonas Löwgren, and Erik Stolterman, *Thoughtful Interaction Design: A Design Perspective on Information Technology* (Cambridge, MA: Mit Press, 2004)

Sharon Macdonald, and Paul Basu, *Exhibition Experiments* (John Wiley & Sons, 2008)

Anne Rørbæk Olesen, 'For the Sake of Technology? The Role of Technology Views in Funding and Designing Digital Museum Communication', *Museum Management and Curatorship*, 31.3 (2016), 283–298

Anne Rørbæk Olesen, 'Mapping Innovation Processes: Visual Techniques for Opening and Presenting the Black Box of Service Innovation Processes', in *Research Methods in Service Innovation* (Edward Elgar Publishing, 2017) <http://www.e-elgar.com/shop/research-methods-in-service-innovation> [accessed 4 June 2021]

Anne Rørbæk Olesen, Nanna Holdgaard, and Ditte Laursen, 'Challenges of Practicing Digital Imaginaires in Collaborative Museum Design', *CoDesign*, 0.0 (2018), 1–13 <https://doi.org/10.1080/15710882.2018.1539109>

Anne Rørbæk Olesen, Nanna Holdgaard, and Anders Sundnes Løvlie, 'Co-Designing a Co-Design Tool to Strengthen Ideation in Digital Experience Design at Museums', *CoDesign*, 0.0 (2020), 1–16 <https://doi.org/10.1080/15710882.2020.1812668>

Ross Parry, *Recoding the Museum: Digital Heritage and the Technologies of Change* (New York: Routledge, 2007)

Darren Peacock, 'Making Ways for Change: Museums, Disruptive Technologies and Organisational Change', *Museum Management and Curatorship*, 23.4 (2008), 333–351

Elizabeth B.-N. Sanders, and Pieter Jan Stappers, 'Probes, Toolkits and Prototypes: Three Approaches to Making in Codesigning', *CoDesign*, 10.1 (2014), 5–14 <https://doi.org/10.1080/15710882.2014.888183>

Donald A. Schön, *The Reflective Practitioner: How Professionals Think in Action* (Aldershot, UK: Ashgate, 1983)

Carol H. Weiss, 'Nothing as Practical as Good Theory: Exploring Theory-based Evaluation for Comprehensive Community Initiatives for Children and Families', in *New Approaches to Evaluating Community Initiatives: Concepts, Methods, and Contexts*, ed. by James P. Connell, Anne C. Kubisch, Lisbeth B. Schorr, and Carol H. Weiss (Washington DC: Aspen Institute, 1995), pp. 65–92

Richard Wetzel, Tom Rodden, and Steve Benford, 'Developing Ideation Cards for Mixed Reality Game Design', *Transactions of the Digital Games Research Association*, 3.2 (2017) <https://doi.org/10.26503/todigra.v3i2.73>

About the Authors

Anne Rørbæk Olesen, Christian Hviid Mortensen, Anders Sundnes Løvlie

Anne Rørbæk Olesen earned her PhD in Communication from Roskilde University in 2015 with the thesis *Co-designing Digital Museum Communication*. She has co-authored publications in internationally renowned anthologies and journals, such as *The Routledge Handbook of Museums, Media and Communication*, *Museum Management and Curatorship* and *Museum & Society*. Her research focuses on digital communication, design practices, collaboration and experience design.

Christian Hviid Mortensen, PhD, is a former curator of media heritage at the Media Museum in Denmark (2007-2018). Currently Christian is a postdoc at the IT University of Copenhagen, where he researches the dynamics of hybrid heritage experiences spanning the digital and physical realms. His research interest is the dynamics between media, culture, heritage and memory. Christian is also on the editorial board of MedieKultur: Journal of Media and Communication Research.

Anders Sundnes Løvlie does research on the intersection of design research and media studies, focusing in particular on experience design, locative media and play. Anders was the coordinator for the GIFT project, and has been involved in a number of design projects involving museums like the Munch Museum, Danish Architecture Center, the National Gallery of Denmark, Brighton Museum and the Frederiksberg Museums. Author of the book *The Rhetoric of Persuasive Games: Freedom and Discipline in America's Army.* Leads the research group on Media, Art and Design (MAD).

9. Data-Driven Visiting Experiences

Steve Benford, Dimitri Darzentas, Edgar Bodiaj, Paul Tennent, Sarah Martindale, Harriet Cameron and Velvet Spors.

Abstract

Data is seen as the 'new oil' that drives the digital economy, and museums are no exception to this. We investigate how data captured from hybrid museum experiences can become a resource for designers, for museums, and for visitors in their understanding of a museum and a museum visit. We ground our discussion in three complementary case studies: Analysing visitors' movements in a touring virtual reality exhibition called *Thresholds*; analysing their gifting behaviours in the *Gift* app; and capturing data about how designers used our Visitor Box cards.

Keywords: Data visualization; Visitor behaviour; Data; Digital tools

Data is seen as the 'new oil' that drives the digital economy. This is most obviously true of digitally native products such as the social media platforms and search engines that we use every day which are fuelled by both the data we directly provide (our stories, comments, photos, videos and likes) as well as the data that is implicitly captured about our histories of searching, browsing and viewing. It is also true however of traditional physical products that are increasingly associated with data capture, from smart homes to software-controlled cars and many others. Such data provides rich opportunities to learn about, redesign and ultimately personalize the user experience, as well as to advertise further ones. It has also raised extensive ethical concerns about privacy, trust, bias and other potential misuses of personal data that we revisit below.

Museums are also interested in such data. They already routinely capture data about visitor numbers and ticketing and there have been various attempts to capture richer data about patterns of movement through the museum, dwell times and preferences. The spread of hybrid visiting

Waern, A. and A. Sundnes Løvlie (eds.), *Hybrid Museum Experiences: Theory and Design.* Taylor & Francis Group, 2022
DOI 10.5117/9789463726443_CH09

experiences of the kinds described in this book raises the opportunity to capture far richer data about visitors' behaviours and, as it turns out, designers' thinking too.

In this chapter, we investigate how data captured from hybrid museum experiences can become a resource for designers, for museums, and for visitors in their understanding of a museum and a museum visit. We ground our discussion in three complementary case studies: Analysing visitors' movements in a touring virtual reality exhibition called *Thresholds*; analysing their gifting behaviours in the *Gift* app (see Chapter 3); and capturing data about how designers used our *Visitor Box* cards (see Chapter 8). Having introduced our case studies, we then reflect on the potential benefits of such data to museum designers, developers, curators but also visitors themselves, as well as the challenges they raise.

Visualizing Visitor Behaviour in Thresholds

For our first case study, we turn to a museum installation that was created by the artist Mat Collishaw in collaboration with the Mixed Reality Laboratory at the University of Nottingham. Technically, *Thresholds* is an example of so-called 'substitutional reality' in which a 3D virtual model is overlaid on a corresponding physical set to deliver the experience of passive haptics in which sensations of physical touch appear to be aligned to digital visual and audio stimuli.[1] We include *Thresholds* here because it provides an interesting first example of analysing data about visitor behaviour at scale as a way of gaining insights into museum experience design. A detailed account of the design and evaluation of *Thresholds* can be found in a journal paper by Tennent and colleagues;[2] the following is a brief summary.[3]

Thresholds recreates the 'Model Room', an exhibition that was staged at King Edward's School (Birmingham, UK) in August 1839 at which Henry Fox Talbot presented a display of 93 'Photogenic Drawings' (photographs). *Thresholds* is an artwork that explores how technology changes our relationship with the world. It comments on how a technical innovation

1 Simeone, Velloso, and Gellersen, 'Substitutional Reality'; Hoffman, 'Physically Touching Virtual Objects Using Tactile Augmentation Enhances the Realism of Virtual Environments'; Insko, 'Passive Haptics Significantly Enhances Virtual Environments'.

2 Tennent and others, 'Thresholds'.

3 Please see this video for an overview of Thresholds: https://www.youtube.com/watch?v=acktp-Wy8Nw.

of photography has led to today's visual culture and simulated realities. Substitutional reality is used to give contemporary audiences access to the previously radical technology of photography. By doing this Collishaw draws a parallel between past and present, both in terms of the thrill of new mediated experiences and in terms of the tensions they provoke. There are historical records of Fox Talbot's concerns about Chartist demonstrations in the Birmingham area at the time of the original exhibition, and in the virtual world of *Thresholds* this rioting can be witnessed taking place outside.

To experience *Thresholds*, each visitor dons a backpack PC and wireless head-mounted display that enables them to explore a room-size virtual reality recreation of the Model Room with up to five other visitors at a time. They are guided into an all-white physical room containing model vitrines and whose walls feature blank outlines of windows, picture frames and other details. Through the headset they see a virtual recreation of the model room as it might have been. This appears to be overlaid onto the physical room so that vision, sound and touch work in synchrony. As a result, they can see and hear Collishaw's recreation of the Model Room but also feel it whenever they reach out to touch a vitrine, lean against a wall or otherwise physically encounter the environment. Walking around, they can directly touch nearly everything, from the vitrines, to the frames of paintings, and can peer out of windows to see and hear angry protesters outside. The one exception is the photographs in the vitrines which cannot be touched; however, they can lift them up in the virtual space by hovering their hand above a vitrine in order to inspect them closely. The fire burning in the grate feels warm, while moths flit around the gaslights, and mice scuttle around the recesses of the room. The other visitors appear as ghostlike auras, conveying a sense of presence but without identifying them or encouraging closer engagement, leading to a shared but still isolated experience. A clock slowly ticks and when six minutes have elapsed, chimes, and they are asked to remove their headset, to find themselves once more in the bright white reality of the physical exhibit.

Thresholds toured widely, and has been exhibited at: Somerset House, London; Birmingham Museum and Art Gallery; Lacock Abbey, Wiltshire, UK; and the National Science and Media Museum Bradford, UK among others. The experience achieved an average throughput of 54 visitors per day across these deployments, peaking at an average of 200 at Somerset House.

Evaluation of the visitor experience drew on the conventional forms of observations and interviews with selected participants, comments captured by museums in visitor books and the like, and also on reviews in the press and on blogs. However, what was more novel and interesting here was the visualization of system logs of visitors' movements in the virtual world which shed additional light onto their behaviours.

We collected 5271 complete data logs of visitors' movements and actions in the virtual world including head position and orientation, hand positions and orientations, and interactions with the virtual photographs. Figure 9.1 (top) presents a heat map of the horizontal positions of all visitors' headsets as seen from above, set against the virtual model. Red shows the most popular locations, orange and yellow the next, green less so, while areas that are not coloured were not visited at all.

This reveals clustering around the door as we might expect as this is the entry and exit point for all visitors. It also shows the vitrines to be popular locations and that visitors tended to stand at their sides rather than their ends, reflecting the orientation of the photographs. They also avoided the relatively busy corridors around the outside of the room and through its centre. Windows were popular locations with many pausing to look out at the riot. The notable gap at the top left is where a static ghost avatar was placed so that invigilators had somewhere to stand safely in the physical room, a tactic that evidently worked well.

41% of these visitors picked up images at least once and visitors spent 2.5% of their time holding objects in total. The heatmap in Figure 9.1 (middle) conveys the relative popularity of images in terms of being picked up suggesting that larger images are more likely to be picked up, perhaps because they are easier to grasp, but also suggesting the images further away from the entrance appeared to be more popular than those near to it. This may be because it takes visitors a few minutes to become familiar with the experience, after which many move to the windows to watch the riot, after which they move along to the end vitrines.

Finally, Figure 9.1 (bottom) provides estimates of the spatial distribution of tracking errors, defined as being reported positions that were either outside the physical constraints of the space or more than 50cm away from the previous recorded point (unlikely to occur with logging at 90 Hz). The visualization shows the last reported 'good' position just before the tracking error occurred. We see the most errors around the entrance – this is to be expected as this was a popular location, is effectively outside the tracking space and we observed visitors adjusting the headset for comfort, or simply grabbing it when the tracking was first established which typically

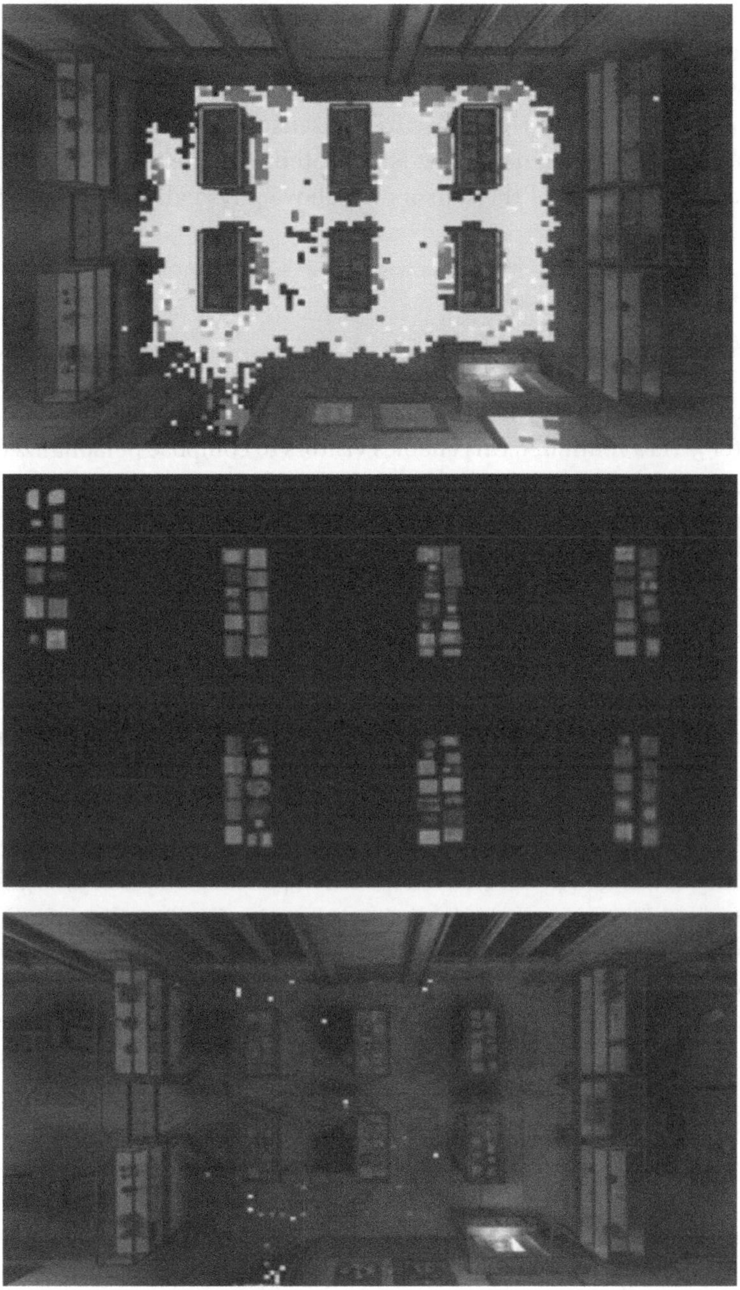

Figure 9.1. Visualizing visitor behaviour in Thresholds. Top: Heatmap of horizontal headset positions in Thresholds. Middle: Relative popularities of photographs for picking up. Bottom: Spatial visualization of estimated tracking errors.

results in their hands covering the sensors. We see several errors around the windows where the mouldings may have obscured the sensors and where we observed users to often physically grab onto and adjust their headphones that might further obscure the sensors. There are also some towards the centre of the room that is consistent with being on the edge of the maximum range of the sensors which we physically placed in its corners.

Visiting Gifting Behaviours

Our second case study turns to the *Gift* app that was described in detail in Chapter 3. As a reminder, this enables visitors to compose personalized museum tours as gifts for others by selecting up to three objects from a museum, taking photos of them and then recording personal messages about why they have chosen each for that person. Initial deployments of the app at the Brighton Museum and Art Gallery and the Munch Museum enabled us to capture data logs about who had sent gifts to whom and what these gifts had contained. In respect of ethical concerns about revealing sensitive personal information, the identities of individuals were anonymized as far as possible and we refrained from any analysis of the contents of their personal messages. In terms of figuring out what they had sent, we needed to manually inspect their photographs in order to determine what they were of; often museum exhibits (though perhaps shot from unusual perspectives), but also sometimes other objects such as selfies, pictures from the cafe, gift shop or even outside.

We developed a series of visualizations of the resulting dataset which comprised several hundred participants and objects with a view to providing insights into visitors' behaviours. The first (Figure 9.2, top) shows what proportions of people who use the app progress through the different stages of the gift-giving workflow. In other words, it shows how many users drop out of using the app at each key touch point of the visitor experience. In this case, we can see that many visitors are lost during the first introductory stage, after which most are retained, though not everyone goes on to include two or even three objects in their gifts (as expected). This is useful for identifying key weaknesses in the overall app or ways in which it is deployed in a particular museum (e.g., are there some aspects of operation that require greater scaffolding from museum staff). It also sets a benchmark as to expected behaviour with the app which might help museums plan deployments and likely uptakes in future deployments.

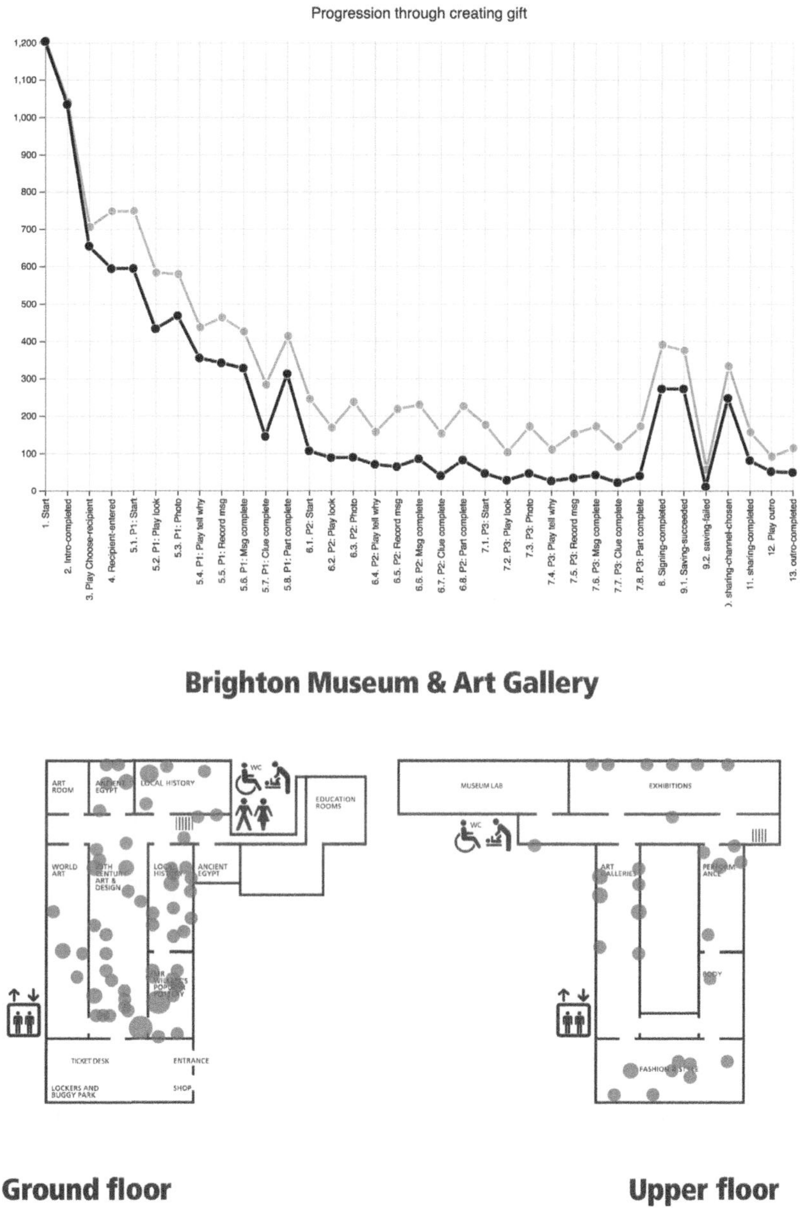

Figure 9.2. Visualizing interactions with the *Gift* app. Top: Visualizing progression through the *Gift* app experience (Green line for Brighton museum data and Pink line for Munch Museum data). Bottom: Map showing the popularity of chosen exhibits for gifting at Brighton Museum and Art Gallery

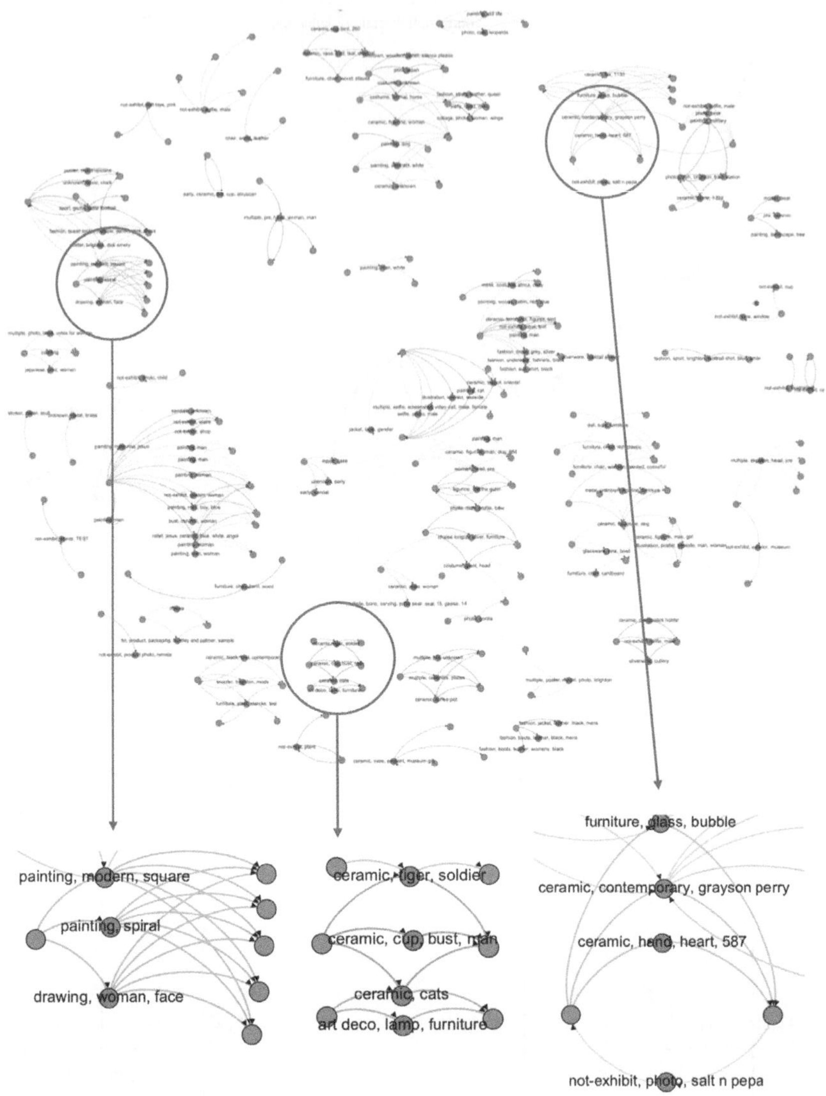

Figure 9.3. Visualizing patterns of gift exchange. Top: Overview of the entire dataset from the Brighton Museum deployment as a Network Graph. Bottom: Zooming in to identify three examples of gifting behaviours.

Our second kind of visualization shows the relative popularity of those museum exhibits that were chosen as gifts, overlaid on a map of the museum. Generating this currently requires further human processing of the data to map the photographs taken by visitors when using the app onto actual exhibits within the museum. Figure 9.2 (bottom) shows this for the Brighton Museum and Art Gallery, revealing how visitors explored the museum widely, often venturing into some of its less frequently visited galleries in search of unusual gifts, which perhaps reflects the relatively non-linear and eclectic nature of its collection.

Our third visualization is an extended form of social network graph. The visualization generates network-style visualizations of gift giving between people, showing who gave which objects to whom. Technically, the visualizations show which devices generated and opened gifts, which is an approximation for people as there is the possibility that some people used multiple devices, or some devices were shared among people. Links from devices to things show whenever the former included the latter in a gift. Links from things to people show whenever the latter opened the former as part of viewing a gift. Figure 9.3 shows an example of such a visualization generated from the Brighton museum data. Figure 9.3 (top) gives an overview of the entire dataset, revealing clusters of gift exchange involving discrete subgroups of participants, and that these appear to involve different patterns of gift giving behaviour in terms of the choice and numbers of museum objects given, how they are combined into gifts, and also the extent to which these are reused (e.g., given to multiple recipients).

Zooming in for a more detailed inspection (Figure 9.3, bottom) reveals several interesting kinds of gift behaviour. Left, we see one person who has made a gift containing three exhibits and then shared it with five others who opened it. Middle, we see three people have made gifts for three others, where their gifts contain several exhibits in common. Right, we see an example of reciprocation between two individuals. Such images suggest the potential to inform our understanding of the social dynamics of gifting museum visits – do some individuals act as 'influencers' perhaps, and is reciprocation a key driver of gifting?

Visualizing the Use of Visitor Box Cards

Our third case study turns to using data to reflect on the design process behind museum experiences rather than on visitors' behaviours. Chapter 8 introduced the Visitor Box deck of cards as a tool for engaging diverse stakeholders in the design of interactive museum experiences. Here we report on a tool called *Cardographer* that was developed to capture data about how cards are used – which are used most often in what combinations by whom – to enable people to reflect in their design processes. This builds on previous work that explored the potential of capturing and analysing data from a deck of Mixed Reality Game ideation cards.[4] The first function of *Cardographer* is to help capture data by using augmented reality technology to identify the presence of the cards during design sessions, and especially on design artefacts.

The second function is then to visualize the resulting data. For example, we captured data from ten design workshops which collectively employed the cards to work up 59 different documented designs. The average number of cards used in a design was 15. The smallest design included 3 cards and the largest design featured 43 individual cards. 3 of the workshops were to train students while the remaining 7 were targeted at museum professionals and accounted for 38 of the designs. An initial analysis revealed considerable variety between the various workshops and hence individual designs with regard to whether they followed and documented all of the stages of the Visitor Box process – only eleven design has all five stages fully documented while some covered only a few stages. This reflects the complexity of the overall Visitor Box process and the time it takes to fully complete it (which may not always fit a short workshop format) as well as the interests of the participants (some may wish to quickly proceed to ideation while others may wish to take their time setting the scene first).

Simply counting the popularity of cards as used can yield some preliminary insights as to their stakeholders' attitudes towards interactive technologies in museums. In this regard, the Visitor Box deck acts as a kind of survey tool to help reveal how the sector is currently thinking. Our data reveals which cards were used most. The following cards were used more than once:

Goals cards reveal overall priorities as: New demographics 12 cards, Use assets in new ways 9, Visitor participation 9, Change visitor attitudes of beliefs 8, Educational activities 4, Visitor numbers 3, Digitize more assets 3, Visitor spend 2, Visitor satisfaction 2, Brand awareness 2, Greater proportion of assets 2.

Motivations cards reveal how they see visitors' motivations for engaging: Curiosity 13, Stimulation 9, Social interaction 7, Academic interest 5, Aesthetic pleasure 5, To make and do 5, Cultural identity 4, Entertainment 3, Time travel 3, Inclusion 3, Wonder 3, Personal relevance 3, Nostalgia 2, Escapism 2, Stimulate the children 2.

4 Darzentas and others, 'Card Mapper: Enabling Data-Driven Reflections on Ideation Cards'.

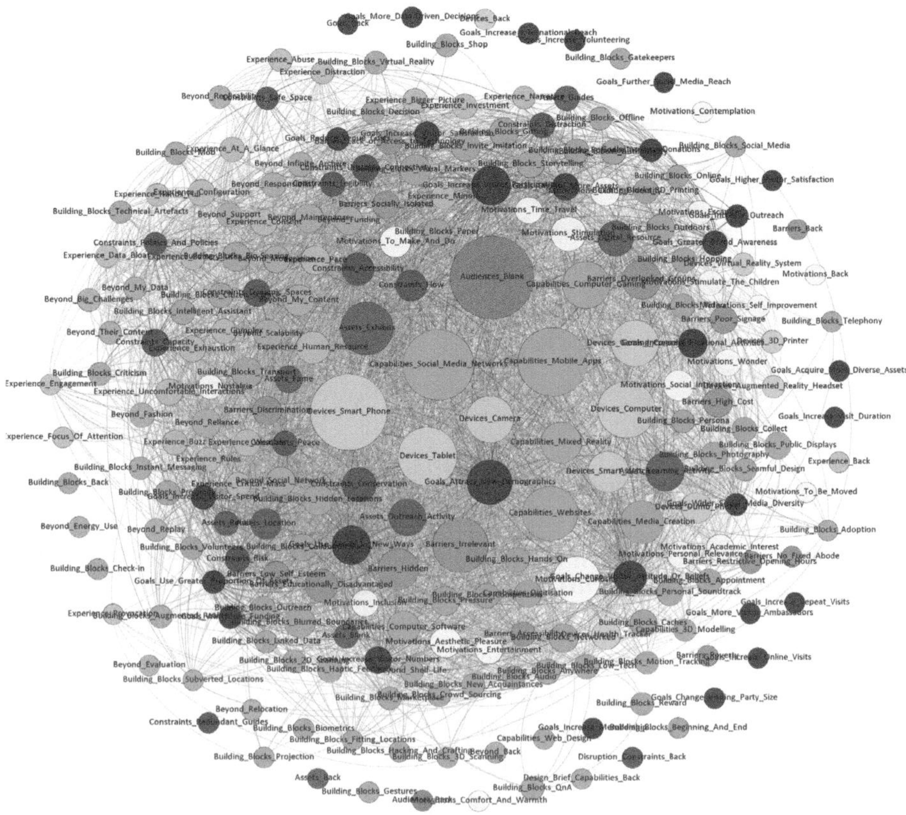

Figure 9.4. The cards perspective

Barriers cards reveal major barriers to digital technology adoption as being: Irrelevant 17, Hidden 10, Overlooked groups 9, Educationally disadvantaged 6, Flow 5, Discrimination 4, High cost 4, Unstable connectivity 3, Low self-esteem 3, Socially isolated 3, Poor signage 3, Accessibility 2, Peace 2, Risk 2, Lack of access to technology 2.

It strikes us that some cards are notable by their absence, not having been used even once. Notable omissions (in our view) include: Visitor Satisfaction, Increase Volunteering, International Reach, To be Moved, and Poverty.

We also explored how our dataset could be further inspected through two complementary visualizations. The first is the Cards Perspective, which gives an overview of all of the cards in the deck and how they have been used as shown in Figure 9.4. This takes the form of a network graph, with each node representing an individual card, the size of which represents the total of how many times this card has been used across all the designs in the

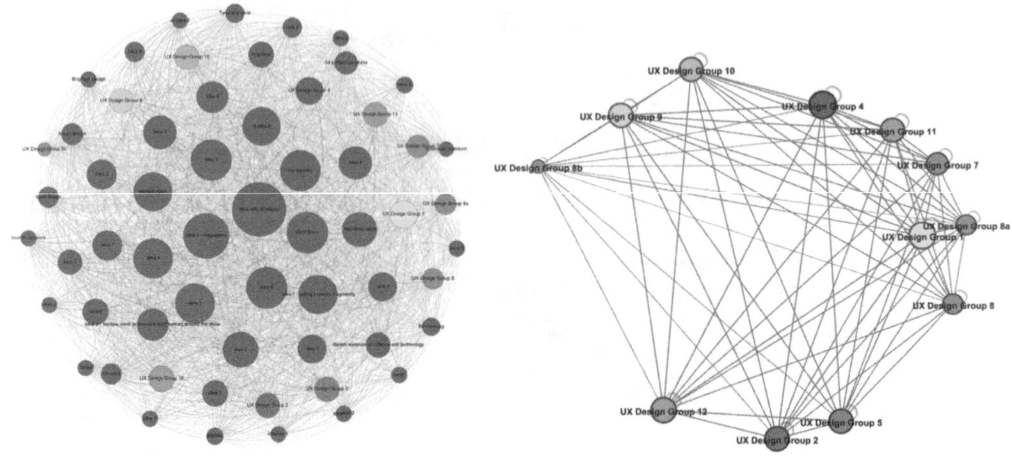

Figure 9.5. The Designs perspective

dataset. The colours match those of the physical cards and convey the theme to which they belong in the deck. Each link between two card nodes denotes the number of times that they have co-occurred in the database of designs (i.e., how often have they appeared together in the same design), with the thickness of the link representing the frequency of these co-occurrences. The card nodes are positioned according to the density of these links. As a result, frequently used and co-occurring cards tend to appear near each other in the centre of the visualization while less frequently used ones appear towards the edge. Cards that are not often used together tend to be positioned far apart. Our visualization is interactive, supporting zooming and panning and also selecting individual cards to show their strength of connection to other cards and data about their use.

The second related visualization is the Designs Perspective as shown in Figure 9.5 left. This shows all of the (currently over 50) designs in the dataset. This is also a network-style graph. Each design that was generated using the cards is shown as a node, with the size representing the number of cards that was involved in generating this design – which can be an early indicator of their complexity. The placement and proximity of the nodes is an indicator of the similarity of the designs in terms of the number of cards they share in common. Two designs are linked if they share at least one card in common. The thickness of the link denotes the number of such shared cards. Again, the placement reveals the centrality or otherwise of designs, at least in terms of the cards they use. Those that use many cards and/or share many in common tend to appear near the centre. Those that just employed a relatively few distinct combinations of cards – which might be a clue as to potentially distinctive thinking – tend to lie towards the edge.

Subsets of these designs will have been generated by a particular organization and/or as part of a particular design process, and it can be illuminating to compare their particular designs to the whole to see how this organization is positioned in terms of its use of the cards. The most detailed case we have of this to date is the use of the cards to teach Masters students at the IT University of Copenhagen (ITU), especially as each was graded by both the course convenor and an independent museum professional. Figure 9.5 right pulls out the ITU students' designs from within all of the designs we captured to help understand their thinking relative to other designers we had encountered. In being shown the visualization, the course convenor was able to reflect that:

> Groups 7, 8, 11 were variations over the same concept, Tinder-style, the user swipes left/right on a bunch of artwork to indicate preferences to be matched with personalized content [...] Groups 2, 5 and 12 were all somewhat conventional designs (in my opinion), and not very original (though different from one another) [...] Groups 4 and 9 were also very similar designs (learning games about interpreting artworks for school students)

Opportunities and Challenges

We now reflect across our three case studies to consider how they highlight various opportunities and challenges for the data-driven design of interactive museum experiences. We consider this topic from three distinct perspectives: Opportunities and challenges for the designers of such experiences, for the curators that host them in museums, and ultimately for the visitors that engage in them.

For Designers

The main opportunity for designers lies in the richer evaluation of experiences, either to inform new ones in the future (i.e., inter-experience evaluation) or to shape an ongoing one as part of an iterative design process (intra-experience evaluation). We saw how in *Thresholds* visualization of fine-grained interactions across thousands of visitors allowed the testing of key design assumptions: Where do people tend to go, look and dwell within an interactive experience and can they be steered to or away from certain areas (e.g., where invigilators stand or areas of poor sensor coverage).

We saw how in *Gift*, one can get a detailed view of the bottlenecks in an experience: At what 'touch points' (to use a term from service design) in the visitor journey do they disengage, which allows for the refinement of the experience so as to deliver more sustained engagement. We also saw how data visualization can reveal unexpected behaviours that suggest new design opportunities. The social graph of gifting behaviours showed that visitors do not always gift museum artefacts as expected but sometimes take photos of other things; might we build on such behaviours to encourage people to share aspects of the visit beyond only the exhibits as part of a gift?

In turn, the data generated by the *Cardographer* tool enables designers to reflect on their own thinking and potentially to compare themselves to others. Are they drawing on the same ideas time and time again, or perhaps ignoring emerging concepts that might help expand their thinking? In more formal terms, are they suffering from 'design fixation'?[5] At a meta-level, the data can also inform the designers of the Visitor Box cards themselves: Are very popular cards conceptually overloaded and could they be split up into sub-concepts? Are little-used cards redundant or perhaps instead interesting outliers? And can the rules of using cards in design sessions be adapted (e.g., insisting that people choose at least one 'rare' card so as to encourage them to think more laterally)?

A key challenge facing designers lies in harvesting data in the first place. Technically, can they reliably capture it remotely from different museums where an experience is deployed (both *Thresholds* and *Gift* toured to multiple museums), and do they have the infrastructure to manage it? Legally and ethically, what are the appropriate terms under which this might be done, and how are museums and visitors involved?

For Curators

The kinds of data we have presented above suggest opportunities for curators and other museum staff to learn more about their museum and visitors. The *Gift* data, for example, highlights which objects were popular as gifted items and where they were in the museum with potential implications for physical curation: What kinds of objects should be displayed and where? They might also guide the curation of online digital collections: One could prioritize digitizing and recommending on a website objects that people want to gift to each other. There are also obvious commercial implications for refining souvenirs that are available in the gift shop.

5 Crilly and Cardoso, 'Where Next for Research on Fixation, Inspiration and Creativity in Design?'

Another opportunity is to learn more about visitors' behaviours. There have been previous attempts to develop algorithms to identify visiting behaviours from movement data, for example classifying visitors who browse galleries of paintings as being ants, butterflies, fish or grasshoppers and then producing exhibit recommendations based on this approach.[6] The kind of data captured from *Thresholds* might further reveal whether these styles are similarly exhibited in virtual worlds and/or more narrative-driven experiences. We might also be able to segment visitors according to other kinds of behaviour. For example, could further analysis of our gifting graphs reveal the presence of 'super-givers' who enjoy making and giving museum visits as gifts for others and if so, how might the museum support or reward them for being influencers in the network? It might even be possible to profile individual visitors with a view to providing them with personalized experiences. The individual stories associated with gifting might shed light into both givers' and receivers' personal associations with artefacts. However, ethically accessing such data might be extremely challenging as we discuss in the following section.

For Visitors

All too often, personal data appears to be something that is gathered and mined by service providers without the direct involvement of consumers themselves. Even where we understand that data is being captured and can consent to or otherwise control this, it is rarely fed back to us in a way that stimulates reflections and insights into who we are and how we behave. And yet the museum is a place of personal reflection and insight in which engagement with exhibits allows us to understand ourselves better or see ourselves in new ways. How then might visitors engage with their own data?

One option is through souvenirs, using data to generate personalized mementoes of the experience. Indeed, previous research has explored both the generation of tangible data souvenirs from museums as well as co-created photostories from theme parks to which both visitors and the park contributed materials.[7] The kinds of data captured from experiences such as *Threshold* and the *Gift* app might generate souvenirs of various kinds from postcards to tangible gifts.

6 Lykourentzou and others, 'Improving Museum Visitors' Quality of Experience through Intelligent Recommendations'; Sookhanaphibarn and Thawonmas, 'A Movement Data Analysis and Synthesis Tool for Museum Visitors' Behaviors'.

7 Durrant and others, 'Automics: Souvenir Generating Photoware for Theme Parks'; Petrelli and others, 'Tangible Data Souvenirs as a Bridge between a Physical Museum Visit and Online Digital Experience'.

An alternative approach is to directly display the data as labels and signage within the museum to complement traditional exhibit labels and provoke further interpretation. Visitors might be interested in the popularity or otherwise of different exhibits as gifts and also in the personal stories that others have told about them. However, this once again leads us back to issues of data ethics involving both ownership and privacy. The multi-faceted nature of some of our data makes this a challenging question. In the *Gift* app, for example, it is one thing to visualize patterns of gifting among anonymized visitors, but quite another to reveal sensitive details of their highly personal stories. A museum might request a visitor's permission to share, but who would they ask, the giver or the receiver? Who owns the gift and controls how and where it is displayed? Even our apparently anonymized data is fraught with challenges. Consider the case of an individual who makes several distinct gifts for different people where each gift involves unique objects that no one else had given or received. In this case, any of the recipients could spot themselves in the picture by knowing that they had received a particular object. From this, they could then identify the giver and also realise that this person had given many other gifts to other people, which in some circumstances might prove to be an embarrassing revelation. In short, anonymizing this kind of behavioural data is rarely as simple as changing names to numbers, as individuals may be revealed by their distinctive behaviours or connections to others, especially if they are in a minority of some kind. In summary, while this chapter has shown there are great opportunities in data-driven museum design, there are also significant challenges and risks that need to be addressed in tandem before these can be unlocked.

Bibliography

Nathan Crilly, and Carlos Cardoso, 'Where Next for Research on Fixation, Inspiration and Creativity in Design?', *Design Studies*, 50 (2017), 1–38 <https://doi.org/10.1016/j.destud.2017.02.001>

Dimitrios Darzentas, Raphael Velt, Richard Wetzel, Peter J Craigon, Hanne G. Wagner, Lachlan D. Urquhart, and others, 'Card Mapper: Enabling Data-Driven Reflections on Ideation Cards', in *Proceedings of the 2019 CHI Conference on Human Factors in Computing Systems*, 2019, pp. 1–15

Abigail Durrant, Duncan Rowland, David S Kirk, Steve Benford, Joel E. Fischer, and Derek McAuley, 'Automics: Souvenir Generating Photoware for Theme Parks', in *Proceedings of the SIGCHI Conference on Human Factors in Computing Systems* (ACM Press, 2011), pp. 1767–1776

Hunter G. Hoffman, ‘Physically Touching Virtual Objects Using Tactile Augmentation Enhances the Realism of Virtual Environments’, in *Proceedings. IEEE 1998 Virtual Reality Annual International Symposium (Cat. No.98CB36180)* (presented at the Proceedings. IEEE 1998 Virtual Reality Annual International Symposium (Cat. No.98CB36180), Atlanta, GA: IEEE, 1998), pp. 59–63 <https://doi.org/10.1109/VRAIS.1998.658423>

Brent Edward Insko, ‘Passive Haptics Significantly Enhances Virtual Environments’ (unpublished PhD Thesis, The University of North Carolina at Chapel Hill, 2001) <https://dl.acm.org/doi/10.5555/933178>

Ioanna Lykourentzou, Xavier Claude, Yannick Naudet, Eric Tobias, Angeliki Antoniou, George Lepouras, and others, ‘Improving Museum Visitors’ Quality of Experience through Intelligent Recommendations: A Visiting Style-Based Approach’, *Workshop Proceedings of the 9th International Conference on Intelligent Environments*, 2013, 507–518 <https://doi.org/10.3233/978-1-61499-286-8-507>

Daniela Petrelli, Mark T. Marshall, Sinéad O’Brien, Patrick McEntaggart, and Ian Gwilt, ‘Tangible Data Souvenirs as a Bridge between a Physical Museum Visit and Online Digital Experience’, *Personal and Ubiquitous Computing*, 21.2 (2017), 281–295 <https://doi.org/10.1007/s00779-016-0993-x>

Adalberto L. Simeone, Eduardo Velloso, and Hans Gellersen, ‘Substitutional Reality: Using the Physical Environment to Design Virtual Reality Experiences’, in *Proceedings of the 33rd Annual ACM Conference on Human Factors in Computing Systems*, CHI ‘15 (New York, NY, USA: Association for Computing Machinery, 2015), pp. 3307–3316 <https://doi.org/10.1145/2702123.2702389>

Kingkarn Sookhanaphibarn, and Ruck Thawonmas, ‘A Movement Data Analysis and Synthesis Tool for Museum Visitors’ Behaviors’, in *Advances in Multimedia Information Processing – PCM 2009*, ed. by Paisarn Muneesawang, Feng Wu, Itsuo Kumazawa, Athikom Roeksabutr, Mark Liao, and Xiaoou Tang, Lecture Notes in Computer Science (Berlin, Heidelberg: Springer, 2009), pp. 144–154 <https://doi.org/10.1007/978-3-642-10467-1_12>

Paul Tennent, Sarah Martindale, Steve Benford, Dimitrios Darzentas, Pat Brundell, and Mat Collishaw, ‘Thresholds: Embedding Virtual Reality in the Museum’, *Journal on Computing and Cultural Heritage*, 13.2 (2020), 12:1-12:35 <https://doi.org/10.1145/3369394>

About the Authors

Steve Benford, Dimitri Darzentas, Edgar Bodiaj, Paul Tennent, Sarah Martindale, Harriet Cameron, and Velvet Spors.

Steve Benford is the Dunford professor of Computer Science at the Mixed Reality Laboratory at the University of Nottingham where he directs the Horizon 'My Life in Data' Centre for Doctoral Training and the 'Smart Products' beacon. He previously held an EPSRC Dream Fellowship, has been a Visiting Professor at the BBC and was elected to the CHI Academy in 2012. He has collaborated with many artists over the past thirty years to create, tour and study interactive performances and installations with a view to gaining new insights into how humans can experience computers.

Dimitrios Darzentas is a multidisciplinary Research Fellow in the Mixed Reality Lab at the University of Nottingham's school of Computer Science. His background is broadly positioned around *Cultural Informatics, Game Design and Technology, Mixed Reality Storytelling*, and the *Design of Meaningful Things*. He is currently working on several EU and EPSRC funded projects involving the use of technologies such as Virtual and Augmented reality and 3D Scanning in interactive museum experiences. These explore how such technologies can enable end-to-end capture of meaningful content, whether tangible or intangible in nature, and their use through the creation of engaging experiences and exhibits. Beyond these he is also working on projects and evolving fronts involving Future Foods, Circular Economies, Experience and Game Design, and Hybrid Physical-Digital Artefacts.

Edgar Bodiaj is a research associate and developer at the Mixed Reality Lab. He has a background in Computer Science and is currently involved with projects focusing on user interaction and engagement inside Virtual Reality. His research interest includes VR experiences, Data Visualisation, and Novel Online Communication.

Paul Tennent is an assistant professor of mixed reality at the University of Nottingham. He has been working with artists and institutions round the world to deliver and study 'in the wild' experiences for more than a decade. His research is around sensing, sensation and sensitivities, often involving deployments that are physically and/or psychologically uncomfortable.

Sarah Martindale is an audience researcher with an interest in how people attach meaning and value to digital interactions and new media; she is a Nottingham Research Fellow in the Department of Cultural, Media and Visual Studies at the University of Nottingham.

Harriet Cameron is an interdisciplinary PhD researcher at the Mixed Reality Lab, University of Nottingham, based within the Horizon CDT. Her research explores identity, power, space/place, and how technologies interact with and configure these experiential concepts, particularly in galleries and museums.

Velvet Spors is a creative technologist and doctoral researcher at the Mixed Reality Lab, University of Nottingham, based within the Horizon CDT. They are currently investigating digital self-care technologies in a public setting (in partnership with the National Videogame Museum). Their research foci are empathy and care in/with/through technology, games and (implicit) interconnectedness between people.

10. Evaluation

Jon Back and Jocelyn Spence

Abstract

This chapter gives an overview of methods and tools for evaluating hybrid experiences in a museum context, and in general, what is gained by doing studies of visitor experiences. It offers strategies for the why, what, who, where, when, and how of conducting evaluations. This includes goal-setting for multiple stakeholders, formative studies, analyses, and ethics. The strategies cover both quick-and-dirty methods as well as in-depth studies.

Keywords: Evaluation; Questionnaires; Interviews; Observations; Data logs; Analysis

Evaluation refers to the process of studying a design with the goal of understanding if it does what the designer team expected it to do. Evaluations can be done as part of the design cycle, experimentally (by temporarily installing an experience), or post-design once an experience is permanently in place in a museum. Evaluations are done through gathering empirical material about how people engage with the museum experience, what they think of it, and what happens in the museum when they engage with it. A very good overview of methods and approaches that have been used for evaluating museum experiences is available in the recent book *Human-Computer Interaction in Museums*.[1] However, most documented evaluations have been done for research purposes. The aim of this chapter is to focus more specifically on hybrid museum experiences, and to present strategies for planning and executing an evaluation that become useful in practice, for the museum and the museum practitioner.

Evaluations tend to happen late in the process, when much of the design work is already done and very little can be changed. In the best of cases, it is

1 Hornecker and Ciolfi, 'Human-Computer Interactions in Museums' Ch. 5.

Waern, A. and A. Sundnes Løvlie (eds.), *Hybrid Museum Experiences: Theory and Design*. Taylor & Francis Group, 2022
DOI 10.5117/9789463726443_CH10

possible to make formative evaluations (informal and early studies) during the design process and use that knowledge to iteratively tweak the design. For example, in the *Your stories* case discussed in Chapter 5, formative evaluations were used to design the clarity of the connection between item and story. But even when formative evaluation is not possible, end evaluations make it possible to learn more about how the design works, for use in communication and marketing, and for informing future design work. Summative evaluations are done at the end of the design process, and focus on studying the final system in use. They can be centred on assessing if the design meets stipulated requirements. An example of a summative evaluation can be found in Chapter 4, where the final study of the *Never let me go* design focuses on understanding how the app was used by the people trying it.

A problem with evaluating hybrid museum experiences, is that these are, by and large, personal and internal to the visitor. Even with the kind of interpersonal experiences that this book advocates, the meaning-making happens between the people involved in ways that representatives for the museum are not privy to. We can never really 'see' what the visitor's full, in-the-moment reactions are. Hence, evaluations must focus on gathering data about their behaviour in the museum, and by asking visitors to describe what they experienced. In museum contexts, we are often evaluating qualitative and personal experiences for which there are no 'right' answers. While well-structured evaluations make it possible for us to step away from our own preconceptions, to try to see the experience through someone else's eyes, we must stay aware that the next individual may approach the experience differently and get something very different out of it.

There are six questions you will typically want an answer to: Why, what, who, where, when and how? In an evaluation, it is important that you not only cover them, but also that you ask them in the right order.

First, Ask Why!

Why are you doing an evaluation at all? To gather information for the funders or the marketing team, to learn more about visitors, to test out a brand-new idea, or to refine a product you think is almost finished? While the answer is probably a combination of reasons, you won't be able to get an answer to everything. Therefore, you need to prioritize your reasons for evaluating.

Be aware that the kind of answers you need to report back to funders might look quite different from the answers you need to know to determine how well the exhibition is liked, and both of these differ greatly from the more design-specific answers you will need during the design process. Not only do different questions need different forms of evaluation; there might also be different reasons for different stakeholders to conduct the evaluation. The reasons for museum project leads and management to conduct a study may include figuring out (for both external and internal stakeholders) what the designers have been up to, especially as hybrid experiences are quite new to many museums. This, in turn, could serve to make internal priorities around audiences, or to gain concrete data to report back to external funders, or to use as a proof that the current work is worth investing in for new external partners. For designers of experiences, common reasons include finding ways to refine the design based on specific areas of success or failure, or to solicit responses on which new ideas can be based.

If you need information for someone else, such as a funder or a marketing team, make sure you know precisely what they are looking for. Whether evaluating an almost finished experience or testing a new design, be clear exactly what it is you hope the design will accomplish. For example, if it is aiming to be an interesting new take on an audio guide, you might want to compare your design with previous evaluations of audio guides, or follow standard methods for evaluating the same.

In this book, there are several examples of evaluations. These were done for research purposes, and in all three cases the 'why' is related to the researchers, and authors of this book, wanting to gain a deeper understanding of hybrid experiences. In Chapter 3, evaluations of the *Gift* app (among other things) strived to develop an understanding how visitors perceived the exhibition and the movement through it, when using the app. Similarly but not the same, in Chapter 5, the evaluation of *Your stories* focused on changes of involvement with the exhibit when using the app. In Chapter 4, *Never let me go*, the focus was rather on understanding what the users would use the app for.

What Is Important to Know?

In general, it is not a good idea to attempt to plan to gather 'everything'. Data collection can be time-consuming as well as expensive, and the more data you have collected the more time it will take to analyse it. Time is an important aspect, both for you and for the visitors. They are not primarily there to help your evaluation, and you must assume they have other interests and obligations. They might not be able to pay full attention to the museum experience. Taking this as a starting point, decide what is most important to know – and what you can influence to guide your participants' attention.

If your marketing team wants you to collect data for them, you will need to include some level of demographic data collection into the data collection even if it is not primarily done for this purpose. If possible, ask only for demographic information that will directly help the marketing team's work. It is often much less important to know whether a visitor was 30-34 or 35-39 than it is to know that the visitor had three small children in tow. Offering a 'prefer not to say' and/or an 'other' option for all demographic questions is good practice, as well as being as inclusive as possible regarding gender, ethnicity, etc.

When preparing a survey or an interview guide you will need to be very careful and precise in the wording of questions. When your list of questions is ready (hint: It isn't yet), share it and discuss it with colleagues to get their input on what is missing, what is unnecessary, and what could be clearer. Let them try to answer the questions. For surveys and interviews, try to make the questions as clear as possible and arrange them so each one builds on the previous ones. Try to combine any overlapping or repetitive questions into one. Use the feedback from colleagues and trials to get down to as few questions as possible, phrased in the clearest possible way, in the most sensible order. A very similar process applies to the development of observation protocols (if you plan to use them): Well in advance of doing observations you will need to discuss what codes to use, and what to observe to get the answers you need, to see what works in practice. Take some time to practice before starting the actual data collection!

Who Should Your Study Subjects Be?

Once you know what questions you want an answer to, you must ask yourself who your evaluation participants will be. What kind of subjects you are looking for, and how you will recruit them, depends on the stage of the process you are at as well as on what questions you have. Some questions might be better answered by a wide, representative sample, while other questions might be more sensitive to having the right audience.

Recruiting participants can be one of the greatest obstacles towards achieving a satisfactory evaluation. Online recruiting might be easy, but it excludes people who are not online much, and social media tends to favour existing audiences (which may or may not suit your purposes). Another option is to recruit on-site at the museum by asking any interested visitors. This risks excluding people visiting with small children, people with limited time to spend, people who are naturally shy, or anyone who does not feel entitled to speak their mind to a museum professional. The latter may happen if questions are asked in a language they do not feel confident speaking, when reaching out to people who do not often visit cultural heritage institutions, or in general to those who tend to feel marginalized for any reason. Do not underestimate the level of initiative it takes for many people to try something new, and then invest their time and energy in an evaluation. If you want to reach new audiences, go where they go, use their language, and be sensitive to their needs.

If you want to know what 'real visitors' think in a 'real-world' context, you need to include as much of the authentic 'onboarding' process as possible. In other words, if they would need to download an app or use a VR headset to use your design, don't dive straight into the design. See how they manage the download process, or how well they can manage the headset on their own. This involves balancing your need to get a useful amount of data, against the desire to find out whether people will engage without any prompting at all. One option is to reach out to museum stakeholders (volunteers, donors, people signed up to your mailing list, people linked to you on social media, etc.) and ask them to sign up for a particular time slot in advance. This will sacrifice some of the 'real-world' context, but it will guarantee a minimum number of responses to a situation that is very near to the pure 'real-world' context. Anyone who takes part beyond these invited participants is then a welcome addition.

Where Will You Be?

Where will you conduct your evaluation? Will you need any extra space, either inside or aside from the exhibition? This might include a space to store equipment, chargers and extra memory cards for video cameras, and to relax for a while between groups. If you are conducting interviews, you will probably need a quiet room for holding those interviews. If they come in groups, will you interview them together, and will there be room for them all to sit comfortably? Will you be asking potentially sensitive questions that might benefit from one-on-one interviews in a private space? If your institution has a café, will they allow interviewers to take up valuable table space for extended periods of time, and/or will it be too loud to record the interview?

Where will people wait? Are there seating, bathroom, and refreshment facilities nearby, especially for people of varying abilities? Might weather pose a problem? The answers to these and similar questions will determine how many subjects you can have in your study, and maybe even the number and types of questions you will be able to ask.

When and for How Long Can You Study Them?

Do not expect much patience from anyone. If participants like the experience, they may spend time on it, but it will be hard to keep them engaged for any extended amount of time afterwards. Ten to fifteen minutes is a good guideline for the maximum length of time to expect an unencumbered visitor who is not pressed for time to commit to an interview, or fill out a form after visiting the experience. Twenty to thirty minutes is a very generous amount of time for interviewees to give, and anything over thirty minutes will exhaust all but the most talkative and opinionated. Take into account as many factors as you can, based on what you know about your visitors and their situations. For one extreme example, if you are targeting school groups, you will need lots of advance planning, the consent of school administrators, teacher and parental permission, and in some locations legal permission for interviewers to speak with children on their own.

Your busiest days of the week and times of day are likely to be your busiest days and times for interviews, as well. This will require thoughtful advance planning.

Another question of 'when' is, when in the design process should you make your study? This depends on your 'why'. If you are evaluating to see how well your experience works, you will study it once it is in place, but if you are trying to inform your choices in the design process, you will need to evaluate during the design process. We will return to this at the end of the chapter.

How to Get Your Answer

Once you know your questions, and you have an idea of whom, where and when you will study, you finally have enough knowledge to decide on your method. The best way to get answers to your questions depends on what your questions are. If you want to know how the experience, or your app, felt, you might want to ask your subjects in an interview. If you are more interested in how people move through the exhibit and how long they stay engaged with your experience, observations and data logs might be better. All methods have their own strengths and weaknesses, and you need to understand the tools you decide to use. While there are many methods, with different epistemological underpinnings that could be delved into, here we take a more practical approach, presenting the most common and useful for evaluating hybrid experiences.[2] From there you can make your choice of which one, or which combination you will use. No matter what method you decide on, it is important to also plan the practical details. How will you make sure that all data ends up saved in the right location? How will you make sure cameras and recorders don't run out of battery? How will you make your schedule so that everybody will have time for breaks and lunch? The better prepared you are, the easier and less stressful it will be on location. Consider also the legal and ethical implications of gathering, managing, and saving a large amount of data where visitors are often identifiable (more on this below).

Deep and Narrow: Interviews

One extremely useful evaluation tool is a one-on-one, in-person interview, immediately after the visitor uses the designed experience. Specifically, we recommend what is known as a semi-structured interview, which simply means that you write out the interview questions beforehand but you don't restrict yourself to using *only* those questions. In this way, you make sure that all participants will answer all questions, but the order may depend on what the interviewee brings up, and if an interviewee says something unexpected, the interviewer can follow up with further questions in a natural, conversational manner. Since all interviewees still answer the same core questions that were decided on beforehand, you can safely make statements such as '80% of our interviewees agreed that the interface was simple to use', but you can also get rich detail on responses you had never anticipated. In fact, if you notice several interviewees mentioning the same unexpected type of response early on, you can adapt your 'interview schedule' (the written list of questions to ask everyone), to include a question about this newly discovered phenomenon.

If an answer is too short to be useful, don't be afraid to follow up with a question such as 'could you tell me more about what you mean by that?' Expect most interviewees to be polite. Many will say that something is

2 Creswell, *Qualitative Inquiry and Research Design.*

'good' or 'nice' or 'interesting' when really, they also have criticisms. Start every interview with the reassurance that you (the interviewer) really need to hear all feedback, good and bad. If possible, tell them at the beginning that you (the interviewer) did not design or build the experience you are asking them about, so you won't be offended to hear criticism.

Compounding the 'too polite' problem is that many people will assume that any problems they had were their own fault. They will be hesitant to admit their struggles for fear of seeming stupid. If you suspect this to be the case, assure them that the goal of the design is to be easy to use, so any problems are the fault of the design. If you suspect that they struggled to understand an unusual concept, reassure them that they are not the only person to find it odd or difficult. Even if they really were doing something 'wrong' that no one else had a problem with, there is much more to be gained by making them feel confident than in establishing who is 'correct'.

If you are interested in group dynamics, you may want to interview groups together. If so, make practical arrangements for the whole group to reconvene at a particular time and place rather than letting the quickest one wait and forcing everyone to find the straggler at the end. Be aware that the dominant speaker in the group is likely to steer the conversation, sometimes so much that you may never hear the full range of opinions without directly asking the quieter participants. Balance this concern against the time and effort on your side of conducting simultaneous individual interviews.

Audio recordings are often the best choice for making the interview go quickly and easily for your interviewees. Plus, if something interesting comes up in a later interview, you can always listen again to see if earlier interviewees mentioned something similar. The downside is the time spent after the fact, either transcribing the interviews or at least listening through them and taking notes. It is possible to take notes during the interview without making audio recordings, but this slows down the process (for interviewees in particular), and you lose the chance to listen back if the interviewer forgets exactly what they meant, or if you later think of something interesting that did not seem important enough to write down at the time. If you use recordings, make sure to ask for permission to do so.

Always, always use two recording devices in case one fails, and always, always have plenty of extra batteries and/or chargers! High quality audio is not important for voice recordings.

Interviews via telephone or video call (Zoom, Skype, FaceTime, etc.) are acceptable unless they take place more than a few days after the experience you are interviewing them about. It also gives people another opportunity not to follow through – and even the best-intentioned visitor has higher priorities on their to-do list than your interview. Phone interviews are particularly useful for longitudinal studies (intentionally following up over extended periods of time), or if you are testing an online system for distant users, but should otherwise be used only as an emergency solution.

Make sure your interviewers (including yourself) know exactly what to do. Explain practicalities like checking battery levels on both audio recorders before pressing 'record', stating the interviewee's name (or other identification) at the beginning of the recording, making sure the interviewer can be heard as well as the interviewee, differentiating between interviewees in group scenarios, and pressing 'stop' at the end of each interview even if someone else is ready to go (so that you can keep track of one audio file per interviewee). Also explain what you're most interested in learning about. If possible, have interviewers use the experience you are testing and then conduct the interview on them yourself.

Interviewers will be busy devoting their attention to the interviewees who are generously giving their time and energy to your project. Make sure you have someone else available to manage the flow of people during a rush, if your experience is open to multiple evaluations at the same time. It only takes one waiting visitor to turn a relaxed interview process into a stressful encounter. Do not risk your reputation by making an engaged, invested visitor feel annoyed, uncomfortable, or unappreciated.

As it Happens: Observations

It is possible, if you have the human and/or tech resources available, to conduct observations of visitor behaviour. Unlike interviews, in observations you will see what actually happened, rather than how the participants remember it. On the other hand, unlike interviews, it is a completely external view capturing what people did, not why they did it. This difference is important to take into account when deciding what method to use to get answers to your specific questions.

Observations can be done by having people in the museum space making detailed notes about whichever behaviours you are most interested in, or it can be done through recordings and later transcriptions of videos. Both work best for capturing obvious gestures around large interfaces and installations that can be viewed from a distance. This is for two main reasons: First, people will change their behaviour if they know or suspect that they are being watched, even if only subconsciously. Second, many visitors are likely to feel very negative about the possibility of being watched or recorded. Assuring them that you are not recording their name or other identifying information may not counter this negative feeling. It is important to handle the situation with care. You do not want to risk your institution's reputation for the sake of an evaluation!

One example where we used observational data in GIFT was to help evaluate the *Gift* app (see Chapter 3). We were interested in seeing how long visitors spent looking at various objects in the museum collection compared to how long they spent looking at the app on their smartphones. We also wanted to see whether the app seemed to bring groups or pairs closer together or drive them towards solitary experiences. We did not have enough people to observe every room in the museum, so we focused on the main gallery. Conveniently, it has a very high, shallow mezzanine running along the shorter walls. Two or three researchers at a time sat quietly in one corner where they could see every visitor as they started using the app, and most began with the main gallery. By dividing up the roughly six to twelve app-testing visitors who would be in the main gallery at the same time among two or three researchers, we were able to keep track of their behaviours quite accurately. We were able to judge the time spent to the nearest quarter of a minute. We could also see the entire gallery except for a few exhibits hidden behind large vitrines and the few directly below our position. Most importantly, no one paid us the slightest bit of attention, with those very few who looked up in our direction perhaps taking us for art students sketching the gallery. However, from that distance, we had no chance of observing what part of the phone app each user was seeing, or hearing, at any given time.

Shallow and Wide: Data Logs

Data logs derived from digital devices are pure gold in terms of ease of collection. Especially in a physical-digital hybrid experience, chances are that any digital element in your design collects data about how it is being used. If you are developing your own, you can decide what information you want it to collect and make sure it does so. Examples include start and end times for using an app, what device it is being installed on, as well as the times at which users initiate any action. This can indicate how long they engage with a particular section of the experience, how long they take performing an interaction, which choices they make, whether they do something 'wrong' or go astray from what was intended, etc. User-generated data such as photos or comments can also be captured for later analysis.

Depending on what exactly you are using, the information that you want to log may be more or less difficult to access. Data is also likely to be presented in a form that is difficult to make sense of. Sometimes, it is simply a case of exporting the data to be imported into a spreadsheet (.xls, .xlsx, etc.). Sometimes the export option is a .csv file – Comma Separated Values – which are also easily imported into spreadsheets. Anything else might require someone with some level of technical expertise to turn automatically generated data into something a person can make sense of.

Exporting the data is just a first step towards making sense of it, as discussed in Chapter 9. What information will actually be useful to you? More is not always better! Depending on the technology you are using, you might pare down your data when deciding what information in those logs to export, or when facing your spreadsheet for the first time. Ideally, you will be able to discuss the possibilities with a developer who can think of new and helpful information to gather, and/or you can prevent them from spending time and energy building in ways to collect data that will ultimately be useless to you.

In terms of outputs, the information generated by data logs is exceptionally useful for very large data sets (in other words, more nuggets of information than you would get from interviewing several tens of individuals). It is also wonderful for correlating different types of information that you would not instinctively assume had any connection. Even without a degree in statistics, graphical outputs such as charts and tables can be extremely helpful for understanding and communicating the results of your evaluations to external stakeholders. A picture is worth a thousand words, as they say – and it certainly has less of a risk of annoying visitors you would otherwise be pestering for an interview.

Data logs can be helpful even when you are interviewing people about their experience. Their honest estimates of time spent on different activities can sometimes be wildly incorrect due to a host of subconscious reasons, whereas data logs are accurate down to fractions of a second. Be aware, though, that the amount of time it took someone to perform an interaction does not necessarily mean that they were focussed on that interaction for the entire time logged – especially in a museum, where one would hope that they would be distracted from the digital event by the museum collection, the building, the people they are visiting with, other museum visitors, and even getting lost in their own thoughts.

Data logs can also give a reliable indicator of where an interface 'went wrong' for a user. Data logs will reveal when, where, and how they went astray from what they were trying to do (or what you were expecting them to do). As with time spent, though, it is not always safe to assume that everything you or your developers think of as an error was truly an error.

Perhaps the visitor bumped the interface unintentionally or tried on purpose to discover a different approach to their experience that your interface did not support. The most reliable use of this kind of information is to substantiate and investigate problems that you have identified in another way or have reason to suspect based on your own experience with the technology.

A Middle Way: Questionnaires

Questionnaires very often constitute our main way of gathering feedback. They are easy to arrange and present less of a burden on visitors: You don't need to find a room for doing interviews, and they take less time for respondents to answer. Questionnaires can also be distributed both on-line and on-site, although the latter is typically more successful in obtaining answers. Questionnaires tend to only get answers that you are expecting, and thus require that you already have a pretty good idea of what you are looking for. Hence, make sure to include some open-ended questions as a way to potentially capture also that which is unexpected. Below, we present a basic questionnaire structure that can be tailored to the precise needs of your study.

Introduction: This is a short text informing about the purpose of the feedback form and why the visitor has been asked to fill it in. Emphasize that this is voluntary.

Demographics: This section collects information about, for example, the respondent's age, home country, gender, and level of museum experience. It can be placed first or last. If it is short, it is better to have it first, but if this is a long section with many optional questions (which it seldom should be), it is better to place it at the end of the survey. Keep this part short; don't ask for information that you won't be using in your analysis. Consider keeping track of which day and time the survey was filled in.

Be considerate of issues related to privacy. Keep the questionnaire as anonymous as possible (exclude name as well as anything else that can serve to identify the respondent). If you are not doing an anonymous survey, you should still make it optional to fill in information that can identify the respondent or can be considered sensitive, such as name and gender. If you are not doing an anonymous survey, also make sure to follow relevant legal regulations and any other personal data directives for your location (such as Europe's GDPR legislation). Among other things, the latter involves telling your participants how you will handle their data, why the survey is being done, and how long their data will be kept.

Short questions: These are your top-priority questions. For these, it often makes sense to use yes/no questions, scaled questions (such as Likert scales), and/or mutually exclusive checkbox questions. These questions should be easy to respond to, they can be made compulsory if delivered online, and they do not expose the identity of the respondent. If you are doing regular evaluations of different exhibitions, standardize these questions as much as possible to permit comparison.

Full text questions: These are the questions that encourage players to review their experience with your design. Most visitors will enjoy answering these, if you give them plenty of flexibility. Give them the space to write long text here, but do not make them feel obligated to. Ask each question separately. If your questionnaire is online, make it optional to answer these questions! Use questions that encourage some concrete feedback, rather than just rating the exhibition as 'good', such as the pair of 'What part of the experience stands out in your mind?' and 'What worked less well?'

Final open question: End with a final open question such as 'Is there something we forgot to ask?' or 'Do you have any final comments?'

Early On: Formative Studies

Design literature will often emphasize involving users early on in the design process using formative studies.[3] One reason to do so is to better understand who they are and what they do, and more importantly to understand how they *change* their behaviour when the design is in place and in use. This literature will also tend to emphasize how involving users in early studies allows them to have a say about critical design choices. In the museum context, it might be particularly useful to allow future users from target visitor segments to co-design experiences that they themselves would find fun and engaging. Such participatory and user-centred design studies are

3 Ciolfi and others, 'Articulating Co-Design in Museums: Reflections on Two Participatory Processes', XIX; Hornecker and Ciolfi, 'Human-Computer Interactions in Museums' Ch. 5.

called *formative*, and they are typically qualitative, shorter, and involve smaller numbers of participants than a final evaluation does.

The problem with involving users early is that since the experience is yet to be fully designed and implemented, they will engage with something *other* than the experience you are trying to design. Depending on what information you are looking for, you must thus first design the way in which you plan users to be involved. Very early in the process, brainstorming and bodystorming exercises present useful methods to tap into the knowledge and creativity that participants bring with them.[4] There also exists a range of methods that work well within teams of developers with or without user involvement, including the ones discussed in Chapter 8, and more can be gathered from the references. But since these methods are very different from user evaluations, they are not discussed further in this chapter.

However, once a first design has been developed, it makes sense to trial it early, even before it is fully implemented. The goals of such trials are typically twofold. Designers will be looking to 'debug' their design, e.g. in terms of clarifying interfaces that are difficult to understand or adding necessary help texts. They might also want to trial specific design features that they are not sure will work or be appreciated. Museums, on the other hand, will want to learn something about how the experience will fit into the overall museum offer, e.g. what visitor groups it will attract and how long people will engage with it. Hence, it makes sense to stage formative studies as realistically as possible, even at an early stage. It is worth using an early prototype, or even one that is partly or fully simulated.

The data from formative studies can be gathered and analysed in much the same way as for the final evaluations discussed earlier. However, it is important to not assume that the results from early studies can be compared to final evaluations. First, since the design is likely to change after these early studies, people may react very differently to the final design. Also, if designers are using the early studies to debug their design, they may even do changes to their design *during* the studies. There is also a large risk of getting overly positive user feedback, as users are aware that this is a first trial. It is not uncommon for visitors to evaluate what they *think* the final design will look like, rather than giving an honest judgement of the current design.

4 Hornecker, 'Creative Idea Exploration Within the Structure of a Guiding Framework'; Oulasvirta, Kurvinen, and Kankainen, 'Understanding Contexts by Being There'; Schleicher, Jones, and Kachur, 'Bodystorming as Embodied Designing'.

Analyse Your Evaluation

Once you have conducted your evaluation, you need to analyse it. Depending on what you plan to do with the information, you will follow more or less rigorous methods. Academics will need to identify, choose, justify, and follow known methods. They may use highly specialized software and follow multistage processes, with multiple analysts checking their coding against each others' work. Nearly the same results can be accomplished using much more relaxed methods, as long as the analysis follows three basic rules.

Know what you're looking for. Group together responses to your key areas of interest in whichever way the participants have responded or you have observed them. You can glean important information from the way people phrase their responses, how long they take to come up with an answer, how long they looked at a certain exhibit, etc. This will provide much more insight than marking answers as a simple 'agree/disagree'.

Actively look for contradictions and negatives. Even if you fully intend to be fair and unbiased in your analysis, human psychology makes us notice responses that go along with what we hope or expect to see. You need to go look, consciously and conscientiously, for contrary information, which is potentially more valuable than the information you're expecting to find.

Actively look for ideas you never thought of. Especially with hybrid interventions, due to the way they take place in the physical space of the museum, visitors tend to develop a vast array of usages that their designers never imagined. Often the most helpful information to come out of an evaluation is the most surprising. Again, this can be difficult to see unless you make a conscious effort to look from your user's perspective instead of your own.

Ethical Considerations

If you are in academia or partnered with academics, you will almost certainly have to follow the university's ethical procedures. These vary from country to country and even to a certain extent from institution to institution, but you can expect to submit detailed plans for the type of data you collect, how you will collect it, how recruitment will be conducted, how the data will be stored, anonymization processes, and (especially in Europe) the full gamut of GDPR requirements.[5] In many countries, your plans will require approval, usually by committee, before the work can begin, and the data collection process will likely require visitors to read and sign consent forms at the very least. In some countries, consent forms must for legal reasons be long, complex, and off-putting, not to mention confusing for interviewers unfamiliar with them, while in other countries there are readability requirements. The main takeaway is to be as thorough as possible, and to start the preparation process as early as possible.

Even if you have no institutional ethics processes in place, you must follow all legal requirements such as GDPR, plus your institution's own guidelines on how to handle confidential or personal data. If your institution has no such guidelines, look up best practices for places like yours and follow them as best you can. Don't think that you are exempt from data regulations because the interviews are spoken: The minute you save that audio file or start to type up so much as a note on what you learned, you are committing your visitors' personal data to digital processing and storage. Failure to act in good faith with the safety of your visitors' information as your top priority can lead to damage to your reputation and even serious legal repercussions.

We are not trying to scare you away from evaluation! The issues related to privacy and consent are usually easily handled, and negative outcomes are relatively rare. But it would be negligent on our part not to alert you to the need to investigate what is required where you are, and negligent on your part not to treat your visitors as the precious resource they are.

Conclusions

Very often, evaluations confirm what you already have informally gleaned from observing visitors or doing the hybrid experience yourself. But sometimes, an evaluation will show that a design was not quite as successful as you thought, or uncover some unexpected aspect of the experience. If the hybrid experience is already in place, you may want to make changes. But the reason for making an evaluation is mainly to be able to use the knowledge gathered to be able to make better choices and designs *in the future*. Evaluations at the end of the process, evaluating the final system in use, gives us knowledge of what can be made better the next time we build an experience.

5 General Data Protection Regulation, see https://eugdpr.org/.

Bibliography

Luigina Ciolfi, Gabriela Avram, Laura Maye, Nick Dulake, Mark T. Marshall, Dick van Dijk, and others, 'Articulating Co-Design in Museums: Reflections on Two Participatory Processes', in *Proceedings of the 19th ACM Conference on Computer-Supported Cooperative Work & Social Computing* (New York: ACM Press, 2016), XIX, 13–25

John W. Creswell, *Qualitative Inquiry and Research Design: Choosing Among Five Approaches* (Thousand Oaks, CA: SAGE Publications, 2012)

Eva Hornecker, 'Creative Idea Exploration Within the Structure of a Guiding Framework: The Card Brainstorming Game', in *Proceedings of the Fourth International Conference on Tangible, Embedded, and Embodied Interaction*, TEI '10 (New York, NY, USA: ACM, 2010), pp. 101–108 <https://doi.org/10.1145/1709886.1709905>

Eva Hornecker, and Luigina Ciolfi, 'Human-Computer Interactions in Museums', *Synthesis Lectures on Human-Centered Informatics*, 12.2 (2019), i–171 <https://doi.org/10.2200/S00901ED1V01Y201902HCI042>

Antti Oulasvirta, Esko Kurvinen, and Tomi Kankainen, 'Understanding Contexts by Being There: Case Studies in Bodystorming', *Personal and Ubiquitous Computing*, 7.2 (2003), 125–134 <https://doi.org/10.1007/s00779-003-0238-7>

Dennis Schleicher, Peter Jones, and Oksana Kachur, 'Bodystorming as Embodied Designing', *Interactions*, 17.6 (2010), 47–51 <https://doi.org/10.1145/1865245.1865256>

About the Authors

Jon Back is a design researcher at Uppsala University. His focus is on how to create engagement, feelings and experiences in public settings. His work is mainly focused on expanded game formats, where the game reaches out of the computer and into the everyday world. He is highly inspired not only by classic game design, but also by areas such as live action role-play, child's play, storytelling, and street performance. He is a researcher and a practitioner, and besides academic publications he's also published board games and card games, as well as performed playful participatory performance art and organised many kinds of game events.

Jocelyn Spence is an interaction design researcher with an interest in performative experience design, performance, gifting, and cultural heritage; she is a Research Fellow in the Mixed Reality Lab at the University of Nottingham.

Coda

11. Remediating, Reframing and Restaging the Museum

Annika Waern and Anders Sundnes Løvlie

Abstract

The final chapter of the book draws some perspective back to the first chapters, focussing on how hybrid museum experiences may also challenge museums in unforeseen ways. We charter the ways in which a hybrid museum experience can change the way a museum is experienced. Depending on context and the reasons for introducing a hybrid experience, these changes can be seen as design opportunities, but they may also challenge the expectations of stakeholders, including both museum professionals and visitors. In order to make informed design choices in a design project, it is important to be critically aware of these expectations.

Keywords: Change; Museum culture; Technology challenges; Technology opportunities;

Throughout this book, we have argued that hybrid museum experiences carry great potential for engaging audiences in new ways that are deeply meaningful – especially interpersonally. But hybrid museum experiences may also challenge museums in unforeseen ways. For instance, when the location-based game *Pokémon GO* was launched in 2016, the massive numbers of players created new ways to move and interact in public space that also affected museums. Virtual features that were important in the game had been attached to real-world locations using a database of locations of public interest, meaning that many of them were museums – and as a result, many museums now saw increased numbers of visitors who came to play the game. For many museums this meant a welcome opportunity to reach an audience who would otherwise not come to the museum. But in the U.S. Holocaust museum the influx of Pokémon-hunting players became a

Waern, A. and A. Sundnes Løvlie (eds.), *Hybrid Museum Experiences: Theory and Design.* Taylor & Francis Group, 2022
DOI 10.5117/9789463726443_CH11

challenge to the museum's identity, seen by some as an inappropriate use of the museum space.

The example of *Pokémon GO* thus illustrates how using digital media to augment our experience of the physical surroundings carries great potential for engaging audiences in innovative ways; but also, that such hybrid experiences may challenge the existing norms and ways in which we use museums as places of cultural significance.

Museums are in their very nature conservative. The core mission of museums centres on the conservation of heritage for posterity, and in relation to the general audiences, a didactic goal of teaching about this heritage. This is also something of which audiences are well aware. While museum professionals can be very progressive, embracing both new technology and new and playful visitor experiences, they are constrained in these efforts not only by the museum's core mission but also by audience expectations. Hybrid museum experiences thus meet with strong norms and ideals, associated with what a museum can and should be, both among museum professionals and among visitors.

In this final chapter of the book, we charter the ways in which a hybrid museum experience can *change* the way a museum is experienced. Depending on context and the reasons for introducing a hybrid experience, these changes can be seen as design opportunities. But they may also challenge the expectations of stakeholders, including both museum professionals and visitors. In order to make informed design choices in a design project, it is important to be critically aware of these expectations, and work closely with all stakeholder groups in order to strike a balance -- between preservation as a core mission, and change as a necessity in a changing world.

The Ideal Museum Experience

As discussed in Chapter 2, an idea that is often brought up in museum literature is that a visitor experience should be 'transformative'. This ideal is connected with the idea that museums should be 'agents of well-being and vehicles for social change'.[1] Within the GIFT project the goal of creating transformative experiences has been expressed on numerous occasions by museum professionals. Above all, museum professionals wish to avoid *trivialization*. For a curator, a vast web of research and knowledge underlies every decision on what to put on display in the museum, and how. They

1 Silverman, *The Social Work of Museums*, p. 4.

can hardly afford having this work be overshadowed by the introduction of a potentially attractive but superficial experience.

Those museums which have strong, unifying theme in their collections – such as a museum dedicated to the Holocaust – may perhaps find it somewhat easier to offer a deep and meaningful transformative experience than those museums which hold a more disparate collection without a single, clear focus. Many museums also struggle with understanding what kind of transformation they want to facilitate, and have the resources for facilitating. In the GIFT project, we on numerous occasions collaborated with the Museum of Yugoslavia. This museum is dedicated to the history of a dissolved country that ended in a gruesome civil war. The museum is located on the grounds of the former communist leader Josip Broz Tito's palace, and houses the grave of Tito and his wife. Finding the voice of this museum continues to be a struggle for its curators; who understand that in order to create an open-ended transformative experience it needs to encourage critical thinking, or the museum becomes political propaganda. Similar issues have been brought to their edge in the rising controversies around the plans for a WWII museum in Poland, which has become a political battlefield between historians and the country's populist and nationalist government.[2]

The 'transformative experience' can thus be seen as one of those design ideals that rarely is fully realised and furthermore, is fraught with political tension.[3] Hence, while the articulated objective for a hybrid experience may well be some kind of transformative experience, it makes sense to also consider playful, satiric, and ironic designs, and even ways to challenge ideals and norms of the museum visit. After all, not every visitor will subscribe to the museum's cultural capital as one worthy of reverence.

The Aura of the Artefact

What is the essence of the museum? Is it a collection of valuable artefacts, or a forum for ideas? This is a question that cuts through much debate about museum experiences for the last 50 years.[4] The question is important because the idea that physical artefacts are at the core of the museum's identity is still very much present among museum professionals and audiences.

2 Donadio, 'A Museum Becomes a Battlefield Over Poland's History'.

3 Żychlińska and Fontana, 'Museal Games and Emotional Truths'.

4 Cameron, 'The Museum, a Temple or the Forum'; Maleuvre, *Museum Memories: History, Technology, Art.*

Collecting and preserving physical artefacts have been a central tenet of most museums, since the first emergence of museums. The purpose of a visit to the museum is often seen as coming face to face with the artefacts in the collection. In interviews with both museum professionals and museum visitors, we have often encountered this ideal, which sometimes seems to echo Benjamin's concept of the 'aura' as a primary characteristic of artworks.[5] Some museums are responsible for the conservation of certain categories of objects (e.g., historical objects from a certain era and area, or artworks of a particular kind), and are duty-bound by law to collect and preserve objects in this category. When such responsibilities lie at the heart of the museum's identity, it follows naturally that they try to achieve their audience-oriented goals by utilizing the museum's primary asset: Their collection of valuable artefacts.

This ideal also seems to underlie a concern related to digital technology, that is common among museum professionals. This phenomenon is known by the moniker 'the heads-down phenomenon': The fear that visitors will walk through a collection of prized artefacts with their gaze firmly fixed at their smartphone screens, oblivious to the marvellous collections on display around them.[6] This concern has been brought up in literature, and has also been the topic of controversies in social media, such as when a photo of young people seated in front of Rembrandt's 'The Night Watch' while looking at their mobile phones went viral in 2016 (Figure 11.1).[7] Social media users expressed disappointment about the youngsters seemingly favouring their phone screens over the historic artwork behind them, one commenter describing the image as 'a perfect metaphor for our age'. However, according to other people commenting on the photo, the children were in fact using an app provided by the museum as part of the tour, seeking further information about the artwork.

Over-emphasizing the primacy of physical artefacts might be an obstacle to realizing the potential of hybrid experiences. If a hybrid design – such as a smartphone app – succeed in engaging visitors, it is likely to lead to at least some situations like the one illustrated by this social media post. But if this is part of an activity that leads these visitors to engage with the museum's (digitized) collections, seek information, or simply explore in a

5 Sherman, 'Quatremère/Benjamin/Marx'.

6 Lyons, 'Designing Opportunistic User Interfaces to Support a Collaborative Museum Exhibit', 1.

7 Fosh, Benford, and Koleva, 'Supporting Group Coherence in a Museum Visit'; Molloy, 'The Real Story Behind a Viral Rembrandt "Kids on Phones" Photo'; Rennick-egglestone and others, 'Families and Mobile Devices in Museums'.

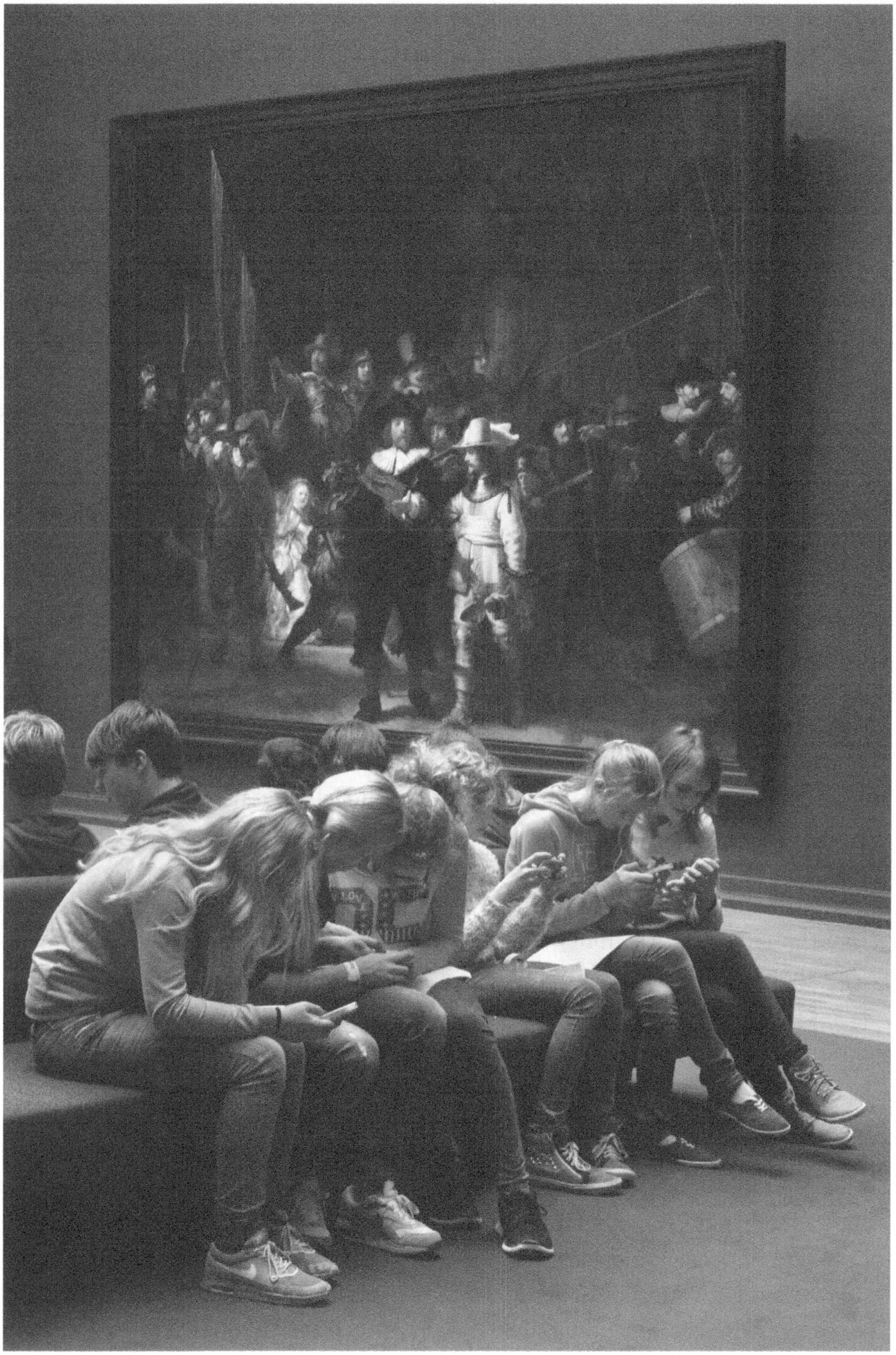

Figure 11.1: Youngsters in the Rijksmuseum. Photo by Gijsbert van der Wal: https://twitter.com/wijdopenogen/status/538085905987567616

new manner, then there is value in this experience that might be lost if we give in to knee-jerk reactions insisting that the digital may not be allowed to take attention away from the physical artefacts.

Reimagining the Museum Experience

As we have seen above, some of the ideals associated with the museum experience are not easily reconcilable with hybrid museum experiences. If we instead turn this perspective around, we might ask: What are the ways in which the digital part of a hybrid experience may change the museum experience? As seen throughout the book, these changes open opportunities for rich, open-ended and interpersonal museum experiences. But they can also be challenging or problematic, depending on how they mix, match and clash with the museum's ideals for a museum visit. Below, we will discuss this in terms of how hybrid design may be used to *remediate*, *reframe*, or *restage* the physical museum experience.

Remediating the Museum

According to one influential theory, new media often tend to imitate formats and genres from older media, a phenomenon referred to as remediation.[8] This should be recognizable to anyone who has ever used a museum 'guide'. In fact, this term has by now become somewhat ambiguous, and could mean many things, e.g. an audio guide, which is usually a device that is rented out by the museum, but nowadays could also be an app that the user opens on their own smartphone; or it could be simply a brochure with information about the museum's highlights – or the term might refer to an actual human guide, as in the original, pre-technological meaning of the word. Over the years, the human guide has been reimagined and replaced through a variety of media, such as printed paper folders, portable audio players, multimedia devices, smartphone apps, and supported through QR codes, indoor positioning technologies, augmented reality, image recognition, etc.

What such efforts above all have in common is that they for the most part adhere quite closely to the established expectations and ideals for a museum experience. For instance, the popular museum app *Smartify* – which boasts of being the most downloaded museum app in the world – offers an experience billed as 'Shazam for art'. When users point their smartphone camera

8 Bolter and Grusin, *Remediation: Understanding New Media.*

at an artwork, the app uses image recognition to identify the artwork and display information about it.[9] While this is a compelling concept built with fairly advanced technology, the main function of the app stays remarkably close to those of the typical museum guide: Providing information about the artwork on the wall. What museum guides have in common, is that they adhere closely to the expectation that museums should *inform* and *educate* visitors about the objects on display. The technology can offer value through increasing the reach and availability of information: Making more information available, making access easier and more flexible, and including media formats that are accessible to larger groups of visitors (e.g., in different languages or in audio rather than text format).

In addition to providing more information, museum curators will often wish for experiences that let visitors engage *deeper* with museum artefacts. Their professional knowledge about the museum objects is vast, and runs deeper than what can meaningfully be put on display. For example, in one of our workshops at the Munch museum one of the curators expressed a desire for the hybrid experience to engage not just with the narrative content of paintings, but with more intricate details such as colour schemas and the choice of textures. Experience designers will have difficulties meeting such requests, as the average museum visitor is not willing to invest the time needed for such a deep exploration – in fact, the average museum visitor typically spends 15-30 seconds looking at a famous work of art (Carbon, 2017). It is an important challenge for designers to craft experiences that can be taken in very quickly, but that still manage to provide a meaningful deepening of the encounter with the museum object. In the GIFT project, we addressed this challenge through a smartphone app called the *One Minute Experience.*[10] This app offers an experience similar to that offered by *Smartify*, using storytelling techniques and an authoring tool that breaks the information into compelling bite-size stories, accessible within a timespan that matches the time visitors might realistically spend in front of the artwork.

Due to the central role of artefacts in the museum experience, hybrid experiences tend to aim to direct the visitor's attention towards the museum artefacts, mostly through associating the virtual content to the individual artefacts. If the experience is intended to keep the visitor's attention focused on objects on display, the interaction that connects the physical object

9 http://smartify.org

10 Løvlie and others, 'Designing for Interpersonal Museum Experiences'; Stoltze, Wray, and Løvlie, 'Reflect in One Minute'.

and the digital content should be designed to support this. This can, e.g. be achieved through augmented reality, by having visitors look at objects through their phone's camera. In this way, visitors can be given a deeper experience of exhibited objects through the use of virtual content, while still keeping the object in direct view. However, this may not always be the best solution – such as, if many visitors gather around particular artefacts and point their phones at those, this may lead to crowding and annoyance to others who would rather perceive the object first hand. One must also keep in mind that museum collections include a host of objects that are not on display. Today, many museum institutions have digitized large parts of their collections. While digital collections provide a rich asset for hybrid museum experiences, any solution focused on digital collections runs the risk of overshadowing the physical encounter with museum objects.

Reframing the Museum

When the hybrid design does not adhere entirely to the traditional expectations towards a museum, the museum experience becomes reframed. This may happen in a number of ways. One approach that we have discussed in Chapter 2 and Chapter 4 is to introduce some kind of game or playful activity into the museum space. Probably, the most common type of game played in museums are variations of scavenger hunts. We typically find examples of scavenger hunts as suggested activities for children, who are tasked with finding particular objects or hunting for specific hidden details in art works. Playing a scavenger hunt in a museum alters the logic of the visit ever so slightly, insofar as the player chooses to go along with it: The game dictates which objects matter (and, by implication, the others do not), and the purpose of the player's movements through the museum is now to find those objects so they can be ticked off on the list, rather than observing and reflecting on the objects for their own intrinsic value.

In Chapter 2, we also brought up that this type of mechanical logic has been criticized as shallow gamification, as it is sometimes seen to ignore the purpose of the activities that are gamified.[11] However, scavenger hunts emerge from a long history of folk games and there is little doubt that they are still a popular activity, and one which obviously lends itself to museums as holders of large collections of unusual objects.[12] Several of the designs in the GIFT project included an element of a scavenger hunt. One example was

11 Deterding and others, 'Gamification: Using Game-design Elements in Non-Gaming Contexts'.
12 Montola, Stenros, and Waern, *Pervasive Games: Theory and Design*, p. 32.

the *Sensitive Pictures* app, which was developed for the Munch museum to experiment with emotions in technology-mediated art experience.[13] This app presented visitors with 6 artworks that they were tasked with finding, in order to hear the stories attached to them.

In Chapter 4, we saw an example of a quite different way of engaging with play in the museum context. *Never Let Me Go* is an example of how a playful app can change the way we perceive a museum through challenging the ways that visitors are expected to reflect on the exhibits as well as how to behave. Encouraging playfulness opens up these opportunities, as well as offers ways to appropriate the hybrid experience itself for creative forms of play. This form of gamification has sometimes been called playification, and is particularly suitable for encouraging social interaction.[14]

In general, hybrid experiences tend to cause people to behave in ways that deviate slightly from the expected norm. In the GIFT design experiments we have asked participants to look down at their phones, search for codes and objects in the museum, sit down in deep thought with closed eyes, stand extended periods in front of paintings listening to a soundtrack, quietly record voice messages, act as mice, etc. The cultural norms of how you are expected to behave in museums are typically perceived as rather strict, and in post-experience interviews, participants have often expressed that they are unsure what behaviours are acceptable in the museum. Participants will even feel uncomfortable about doing things in the museum that they would otherwise willingly do, such as taking a selfie. As exemplified in the introduction to this chapter, visitors often express concern about engaging in playful activities as these are considered to be trivializing. Visitors' perceptions about museum norms, and their confidence in breaking these norms, may also vary depending on their education, socioeconomic background, familiarity with museums (their 'habitus'), etc. Such factors are important for museums looking to reach out to broader audiences.

There are other ways in which a hybrid design might reframe the museum experience. In particular, the central role of museum artefacts may be challenged. In the *Your Stories* installation described in Chapter 5 the museum invited visitors to bring objects that were personally important to them, in order to have them 3D-scanned and digitally 'donated' to the museum. These digital 3D 'artefacts' were made available as augmented reality visualizations located next to historical objects on display in the museum. This design

13 Løvlie and others, 'Designing for Interpersonal Museum Experiences'.

14 Márquez Segura and others, 'Playification: The PhySeEar Case'; Nicholson, 'A RECIPE for Meaningful Gamification'.

challenges the museum collection and curatorial selection of artefacts. What is the difference between the personal value attached to private objects, and the museological value attached to the curated collection of objects on display in the museum? In other design experiments based on 3D scanning, we have experimented with the physical presence of objects in a museum, as typically untouchable and behind glass. We have experimented with using VR technology in combination with 3D printing, to invite visitors to virtually reach into museum cases and bring out the objects (virtually) for closer inspection using both sight and touch.[15]

The museum experience can also be reframed socially. Despite the fact that museum visits are almost always socially negotiated, museum ideals (such as those discussed earlier in this chapter) tend to focus on the relation between the individual visitor and the museum as a cultural institution. This means that even just taking the visitor group into account, and e.g., designing for a family with all their needs and different motives for visiting, can be perceived as a challenge to the ideal museum visit. For example, mobile hybrid experiences might benefit from structuring digital interaction in short bursts, so that the visitor is free to pick up and put away the mobile at any time to focus on the actual exhibition and their visitor group. However, a deep and engaging interaction with the hybrid experience (or even just listening in to the audio guide) may require total focus from its (singular) user.

As exemplified by the *Gift* app (Chapter 3) as well as the *Never Let Me Go* experience (Chapter 4), the social structure of a visiting group can be tapped into to reinforce a hybrid experience. In *Never Let Me Go* the social context is another visitor. The *Gift* app was primarily designed for singular visitors who were tasked with creating a gift for someone who was *not* in the museum. This way, the social context of the visit was reframed, from that of a sole visitor towards a more social experience, even though the gift recipient was not physically present. Both also illustrate how hybrid experiences can be designed to contain no factual information at all, to instead focus on an open-ended, deeply personal and socially meaningful exploration of the museum through tapping into the visitors' social context.

Finally, tapping into and enforcing emotional experiences provides a path towards reframing the museum experience as one less of factual information and learning, towards becoming more deeply and personally meaningful. In one of the Gift designs, the *Sensitive Pictures* project, the

15 Tennent and others, 'Thresholds: Aligning Vision, Sound and Touch in Substitutional Reality'.

design focused entirely on affect.[16] This design was created as a bespoke experience for the Munch museum, an art museum in Oslo housing the life work of Edvard Munch. It explores multiple ways of bringing visitor emotions into play. First of all, visitors were presented with a series of dramatized audio clips designed to explore the emotional themes in Munch's paintings and invite the visitor to reflect on their own emotions in relation to both the artwork and their own life. Second, visitors are invited to share their emotions both by responding to questions in the app, as well as indirectly through technological measurements: EEG sensors and cameras detecting emotions in their faces. At the end of the *Sensitive Pictures* tour, visitors receive a graphical visualization of all the emotional data registered by the system, inviting them to reflect: 'Do you think this is actually what you felt?' As technological developments keep bringing up new ways to capture emotions, and such technologies are increasingly embedded in the products and services we use in our daily lives, there is likely to be an increased interest in exploring this area also among museums and their visitors.

Restaging the Museum

The most radical – and controversial – ways in which hybrid experiences can affect the physical museum are those which threaten to more or less entirely occlude the physical museum, directing the visitor's attention fully towards the digital. As we have seen above it is a common fear among museum professionals that this may happen, and perhaps not entirely without reason. However, it is worthwhile to explore the motivations that might motivate such designs, and the forms they might take.

We noted in the introduction that the popular augmented reality game *Pokémon GO!* led large numbers of players to visit museums, in order to access the virtual resources attached to the museum's location. This might serve as an extreme example of a digital experience that is geographically attached, but topically and culturally *detached* from the museum – potentially setting up crass contrasts with culturally sensitive sites.

In our explorations, we have found that designs that encourage play will sometimes be perceived as a too radical re-staging of the museum. For example, one of our early designs, the *Twitto* game was rejected by museum professionals because it directed the visitors' attention away from the physical objects on display, as well as trivialized their historical significance.[17]

16 Løvlie and others, 'Designing for Interpersonal Museum Experiences'.

17 Back and others, 'GIFT: Hybrid Museum Experiences through Gifting and Play', MMCCXXXV.

While the activities in this game were closely focused on a central topic of the exhibition – the life of Tito – the curators still experienced the design as too uncritical when framed through the lense of gameplay.[18] Similar concerns were voiced by participating visitors.

Is there, then, reasons to sometimes entirely *restage* the museum? It is worth noting that museums and other heritage sites are often used for activities that do not have a direct connection with the museum's collection – such as concerts, performances or parties. It is not hard to find ways in which museums find value in housing activities that are not directly referencing the museum collections. One such example is the staging of larp (live role-playing experiences) and theatre in museums as alternative ways to explore their cultural significance.[19] Hybrid museum experiences can fill similar functions, e.g. for visitors that cannot meaningfully engage with the objects on display. For example, it could be a good idea to design a floor-level treasure hunt for children, leaving their parents to explore the paintings of an art museum. But perhaps more importantly, new museology makes a strong argument to move from stories about objects towards stories about *people*. In an exhibition focused on the history of people, the focus on artefacts may sometimes be replaced with a rich offering of media resources: Audio recordings, video, excerpts from letters. Hybrid museum experiences offer ways in which media can be delivered that are not tied to specific artefacts.

Final Words

Hybrid museum experiences do not just have the potential of creating new visitor experiences; they also bear with them the potential to change the museums where they are made available.

This is both a risk and an opportunity. Introducing hybrid experiences in ways that reframe or restage a museum, risks making it a backdrop to an experience that has little to do with its established role and cultural significance. This, in turn, risks challenging the museum professionals in their status and identity, and very often also challenges the museum visitors' expectations on what a museum visit is supposed to be.

18 Waern and others, 'Sensitizing Scenarios: Sensitizing Designer Teams to Theory'.

19 Bridal, *Exploring Museum Theatre*; Nikonanou and Venieri, 'Museums as Gameworlds: The Use of Live Action Role Playing Games in Greek Museums.'

However, to what extent can museums cling to their established ideals? Museums may be institutions of conservation, but do they really need to be conservative? Very often, the key to a successful hybrid design lies in making clear early what changes it is intended to bring about, and how everyone in a design project – curators, technology designers and museum staff – can work together to bring about those goals. A major part of this book has presented methods to help the project group develop a joint understanding of the underlying expectations, and the often unarticulated ideals, norms and values that the museum embodies. Once those values have been uncovered, it also makes sense to carefully consider what could be gained by challenging them.

It is our hope that this book has provided the theories, examples, and tools needed to make it possible to identify and preserve those museum ideals that are indeed central, while challenging others that are habitual rather than essential, and that might be detrimental for the visitor's engagement or the museum's ability to create transformative experiences.

And now, the continuation is up to you, the reader.

Bibliography

Jon Back, Benjamin Bedwell, Steve Benford, Lina Eklund, Anders Sundnes Løvlie, William Preston, and others, 'GIFT: Hybrid Museum Experiences through Gifting and Play', in *Proceedings of Workshop on Cultural Informatics*, ed. by Angeliki Antoniou and Manolis Wallace (presented at the EUROMED2018, Nicosia, Cyprus: CEUR Workshop Proceedings, 2018), MMCCXXXV, 31–40 <http://ceur-ws.org/Vol-2235/paper4.pdf> [accessed 3 June 2021]

Jay David Bolter, and Richard Grusin, *Remediation: Understanding New Media* (Cambridge, Mass.: The MIT Press, 2000)

Tessa Bridal, *Exploring Museum Theatre* (Walnut creek, Cal: Rowman Altamira, 2004)

Duncan F. Cameron, 'The Museum, a Temple or the Forum', *Curator: The Museum Journal*, 14.1 (1971), 11–24 <https://doi.org/10.1111/j.2151-6952.1971.tb00416.x>

Sebastian Deterding, Miguel Sicart, Lennart Nacke, Kenton O'Hara, and Dan Dixon, 'Gamification. Using Game-design Elements in Non-gaming Contexts', in *CHI '11 Extended Abstracts on Human Factors in Computing Systems*, CHI EA '11 (New York, NY, USA: ACM, 2011), pp. 2425–2428 <https://doi.org/10.1145/1979742.1979575>

Rachel Donadio, 'A Museum Becomes a Battlefield Over Poland's History', *The New York Times* (New York, NY, 9 November 2016) <https://www.nytimes.com/2016/11/10/arts/design/museum-of-the-second-world-war-in-poland-debate.html> [accessed 23 June 2021]

Lesley Fosh, Steve Benford, and Boriana Koleva, 'Supporting Group Coherence in a Museum Visit', in *Proceedings of the 19th ACM Conference on Computer-Supported Cooperative Work & Social Computing*, CSCW '16 (New York, NY, USA: ACM, 2016), pp. 1–12 <https://doi.org/10.1145/2818048.2819970>

Anders Sundnes Løvlie, Lina Eklund, Annika Waern, Karin Ryding, and Paulina Rajkowska, 'Designing for Interpersonal Museum Experiences', in *Museums and the Challenge of Change: Old Institutions in a New World*, ed. by Graham Black (London & New York: Routledge, 2020), pp. 223–238

Leilah Lyons, 'Designing Opportunistic User Interfaces to Support a Collaborative Museum Exhibit', in *Proceedings of the 9th International Conference on Computer Supported Collaborative Learning*, CSCL'09 (Rhodes, Greece: International Society of the Learning Sciences, 2009), I, 375–384

Didier Maleuvre, *Museum Memories: History, Technology, Art* (Stanford University Press, 1999)

Elena Márquez Segura, Annika Waern, Luis Márquez Segura, and David López Recio, 'Playification: The PhySeEar Case', in *Proceedings of the 2016 Annual Symposium on Computer-Human Interaction in Play* (presented at the Proceedings of the 2016 annual symposium on computer-human interaction in play, ACM, 2016), pp. 376–388 <https://doi.org/10.1145/2967934.2968099>

Mark Molloy, 'The Real Story Behind a Viral Rembrandt "Kids on Phones" Photo', 16 January 2016, section News <https://www.telegraph.co.uk/news/newstopics/howaboutthat/12103150/Rembrandt-The-Night-Watch-The-real-story-behind-the-kids-on-phones-photo.html> [accessed 21 February 2020]

Markus Montola, Jaakko Stenros, and Annika Waern, *Pervasive Games: Theory and Design* (Boca Raton, FL: CRC Press, 2009)

Scott Nicholson, 'A RECIPE for Meaningful Gamification', in *Gamification in Education and Business*, ed. by Torsten Reiners and Lincoln C. Wood (Cham: Springer International Publishing, 2015), pp. 1–20 <https://doi.org/10.1007/978-3-319-10208-5_1>

Niki Nikonanou, and Foteini Venieri, 'Museums as Gameworlds: The Use of Live Action Role Playing Games in Greek Museums.', *International Journal of the Inclusive Museum*, 6.3 (2014)

Stefan Rennick-egglestone, Patrick Brundell, Boriana Koleva, Steve Benford, Maria Roussou, and C. Chaffardon, 'Families and Mobile Devices in Museums: Designing for Integrated Experiences', *Journal on Computing and Cultural Heritage*, 9.2 (2016), 11:1-11:13 <https://doi.org/10.1145/2891416>

Daniel J. Sherman, 'Quatremère/Benjamin/Marx: Art Museums, Aura, and Commodity Fetishism', in *Museum Culture: Histories, Discourses, Spectacles*, ed. by Itit Rogoff and Daniel J. Sherman (London: Routledge, 1994), pp. 143–163 <https://doi.org/10.4324/9780203168196>

Lois H. Silverman, *The Social Work of Museums* (London & New York: Routledge, 2010)

Linda Stoltze, Tim Wray, and Anders Sundnes Løvlie, 'Reflect in One Minute', *The GIFT Box* <https://gifting.digital/one-minute-experience/> [accessed 23 June 2021]

Paul Tennent, Sarah Martindale, Steve Benford, Dimitrios Darzentas, Pat Brundell, and Matt Collishaw, 'Thresholds: Aligning Vision, Sound and Touch in Substitutional Reality', *The ACM Journal on Computing and Cultural Heritage (JOCCH)*, Forthcoming (2020)

Annika Waern, Jon Back, Paulina Rajkowska, Jocelyn Spence, Karin B. Johansson, and Anders Sundnes Løvlie, 'Sensitizing Scenarios: Sensitizing Designer Teams to Theory', in *Proceedings of the 2020 CHI Conference on Human Factors in Computing Systems CHI'20* (presented at the Internaltiona, Forthcoming: Association for Computing Machinery, 2020)

Monika Żychlińska, and Erica Fontana, 'Museal Games and Emotional Truths: Creating Polish National Identity at the Warsaw Rising Museum', *East European Politics and Societies*, 30.2 (2016), 235–269 <https://doi.org/10.1177/0888325414566198>

About the Authors

Annika Waern is a 'research through design' academic who has been researching technology-supported physical play and games for about fifteen years. Currently, she is conducting research in the areas of hybrid play in museums, children's play in outdoor settings, and circus training to foster proprioceptic skills. Waern has authored over 100 academic articles and book chapters, and been editor of 2 books and several journal special issues. She is a DiGRA (Digital Games Research Association) distinguished scholar and a HEVGA (Higher Education Video Game Alliance) fellow.

Anders Sundnes Løvlie does research on the intersection of design research and media studies, focusing in particular on experience design, locative media and play. Anders was the coordinator for the GIFT project, and has been involved in a number of design projects involving museums like the Munch Museum, Danish Architecture Center, the National Gallery of Denmark, Brighton Museum and the Frederiksberg Museums. Author of the book *The Rhetoric of Persuasive Games: Freedom and Discipline in America's Army*. Leads the research group on Media, Art and Design (MAD).

Bibliography

Clem Adelman, 'Kurt Lewin and the Origins of Action Research', *Educational Action Research*, 1.1 (1993), 7–24 <https://doi.org/10.1080/0965079930010102>

Michael Ames, 'Thirty-One Propositions on Changing Museums: An Introduction to the Glenbow Case Study', in *Museums and the Paradox of Change*, ed. by R.R. Janes (Calgary, Canada: Routledge, 1997)

Liliana Ardissono, Tsvi Kuflik, and Daniela Petrelli, 'Personalization in Cultural Heritage: The Road Travelled and the One Ahead', *User Modeling and User-Adapted Interaction*, 22.1–2 (2012), 73–99 <https://doi.org/10.1007/s11257-011-9104-x>

Juha Arrasvuori, Marion Boberg, and Hannu Korhonen, 'Understanding Playfulness – An Overview of the Revised Playful Experience (PLEX) Framework', in *7th International Conference on Design and Emotion* (presented at the 7th International Conference on Design and Emotion, Chicago, IL: Design and Emotion Society, 2010) <https://doi.org/10.5281/zenodo.2596777>

Doris Ash, 'Reflective Practice in Action Research: Moving Beyond the "Standard Model"', in *The Reflective Museum Practitioner* (New York: Routledge, 2019), pp. 23–38

Jon Back, Benjamin Bedwell, Steve Benford, Lina Eklund, Anders Sundnes Løvlie, William Preston, and others, 'GIFT: Hybrid Museum Experiences through Gifting and Play', in *Proceedings of Workshop on Cultural Informatics*, ed. by Angeliki Antoniou and Manolis Wallace (presented at the EUROMED2018, Nicosia, Cyprus: CEUR Workshop Proceedings, 2018), MMCCXXXV, 31–40 <http://ceur-ws.org/Vol-2235/paper4.pdf> [accessed 3 June 2021]

Gregory Bateson, 'A Theory of Play and Fantasy', *Psychiatric Research Reports*, 2 (1955), 39–51

Daniel C. Batson, Eklund Håkansson, Jakob, Valerie L. Chermok, Jennifer L. Hoyt, and Biaggio G. Ortiz, 'An Additional Antecedent of Empathic Concern: Valuing the Welfare of the Person in Need', *Journal of Personality and Social Psychology*, 2007, 65–74

Genevieve Bell, 'Making Sense of Museums: The Museum as "Cultural Ecology"', *Intel Labs*, 2002, 1–17

Steve Benford, and Gabriella Giannachi, *Performing Mixed Reality* (Cambridge, MA, London, U.K.: The MIT Press, 2011)

Adam Blatner, *Interactive and Improvisational Drama: Varieties of Applied Theatre and Performance* (iUniverse, 2007)

Linda M. Blud, 'Social Interaction and Learning Among Family Groups Visiting a Museum', *Museum Management and Curatorship*, 9.1 (1990), 43–51

Herbert Blumer, 'What is Wrong with Social Theory?', *American Sociological Review*, 19.1 (1954), 3–10 <https://doi.org/10.2307/2088165>

Fabian Bohnert, and Ingrid Zukerman, 'Non-Intrusive Personalisation of the Museum Experience', in *User Modeling, Adaptation, and Personalization*, ed. by Geert-Jan Houben, Gord McCalla, Fabio Pianesi, and Massimo Zancanaro, Lecture Notes in Computer Science (Berlin, Heidelberg: Springer, 2009), pp. 197–209 <https://doi.org/10.1007/978-3-642-02247-0_20>

Jay David Bolter, and Richard Grusin, *Remediation: Understanding New Media* (Cambridge, Mass.: The MIT Press, 2000)

Inge Bretherton, 'Pretense: The Form and Function of Make-Believe Play', *Developmental Review*, 9.4 (1989), 383–401 <https://psycnet.apa.org/doi/10.1016/0273-2297(89)90036-1>

Tessa Bridal, *Exploring Museum Theatre* (Walnut creek, Cal: Rowman Altamira, 2004)

Richard Buchanan, 'Wicked problems in design thinking', *Design issues*, 8.2 (1992), 5–21

Roger Caillois, *Man, Play and Games* (Champaign, IL: University of Illinois Press, 2001)

Duncan F. Cameron, 'The Museum, a Temple or the Forum', *Curator: The Museum Journal*, 14.1 (1971), 11–24 <https://doi.org/10.1111/j.2151-6952.1971.tb00416.x>

Sebastian Chan, and Aaron Cope, 'Strategies Against Architecture: Interactive Media and Transformative Technology at the Cooper Hewitt, Smithsonian Design Museum', *Curator: The Museum Journal*, 58.3 (2015), 352–368 <https://doi.org/10.1111/cura.12118>

David Cheal, *The Gift Economy* (New York ; Abingdon, Oxon: Routledge, 2015)

Shira Chess, and Adrienne Shaw, 'A Conspiracy of Fishes, or, How We Learned to Stop Worrying About #GamerGate and Embrace Hegemonic Masculinity', *Journal of Broadcasting & Electronic Media*, 59.1 (2015), 208–220 <https://doi.org/10.1080/08838151.2014.999917>

Elise Grimsrud Christensen, 'Working Across Departments to Find Playful and Inclusive Strategies', *The GIFT Box*, 2019 <https://gifting.digital/norwegian-center-for-holocaust-and-minority-studies/> [accessed 2 April 2020]

Luigina Ciolfi, Gabriela Avram, Laura Maye, Nick Dulake, Mark T. Marshall, Dick van Dijk, and others, 'Articulating Co-Design in Museums: Reflections on Two Participatory Processes', in *Proceedings of the 19th ACM Conference on Computer-Supported Cooperative Work & Social Computing* (New York: ACM Press, 2016), XIX, 13–25

James P. Connell, and Anne C. Kubisch, 'Applying a Theory of Change Approach to the Evaluation of Comprehensive Community Initiatives: Progress, Prospects and Problems', in *New Approaches to Evaluating Community Initiatives: Theory, Measurement and Analysis*, ed. by Karen Fullbright-Anderson, Anne C. Kubisch, and James P. Connell, 2 vols. (Washington, DC: The Aspen Institute, 1998), II, 15–44

John W. Creswell, *Qualitative Inquiry and Research Design: Choosing Among Five Approaches* (Thousand Oaks, CA: SAGE Publications, 2012)

Nathan Crilly, and Carlos Cardoso, 'Where Next for Research on Fixation, Inspiration and Creativity in Design?', *Design Studies*, 50 (2017), 1–38 <https://doi.org/10.1016/j.destud.2017.02.001>

Peter Dalsgaard, Kim Halskov, and Rune Nielsen, 'Maps for Design Reflection', *Artifact*, 2.3–4 (2008), 176–189 <https://doi.org/10.1080/17493460802526412>

Areti Damala, Eva Hornecker, Merel J. van der Vaart, Dick Van Dijk, and Ian Ruthven, 'The Loupe: Tangible Augmented Reality for Learning to Look at Ancient Greek Art', *Mediterranean Archaeology and Archaeometry*, 16.5 (2016), 73–85 <https://doi.org/10.5281/zenodo.204970>

Dimitrios Darzentas, Raphael Velt, Richard Wetzel, Peter J. Craigon, Hanne G. Wagner, Lachlan D. Urquhart, and others, 'Card Mapper: Enabling Data-Driven Reflections on Ideation Cards', in *Proceedings of the 2019 CHI Conference on Human Factors in Computing Systems*, CHI '19 (New York, NY, USA: ACM, 2019), p. 571:1-571:15 <https://doi.org/10.1145/3290605.3300801>

Gary Davies, Susan Whelan, Anthony Foley, and Margaret Walsh, 'Gifts and Gifting', *International Journal of Management Reviews*, 12.4 (2010), 413–434 <https://doi.org/10.1111/j.1468-2370.2009.00271.x>

Bernard De Koven, *The Well-Played Game: A Player's Philosophy* (MIT Press, 2013)

Sebastian Deterding, 'Getting Gamification Right' (unpublished Slideshare presented at the Google Tech Talks, Mountain View, California, 2011) <http://www.slideshare.net/dings/meaningful-play-getting-gamification-right> [accessed 5 October 2015]

Sebastian Deterding, Miguel Sicart, Lennart Nacke, Kenton O'Hara, and Dan Dixon, 'Gamification. Using Game-design Elements in Non-gaming Contexts', in *CHI '11 Extended Abstracts on Human Factors in Computing Systems*, CHI EA '11 (New York, NY, USA: ACM, 2011), pp. 2425–2428 <https://doi.org/10.1145/1979742.1979575>

Alan Dix, 'Designing for Appropriation', in *Proceedings of the 21st British HCI Group Annual Conference on People and Computers: HCI... But Not as We Know It* (British Computer Society, 2007), II, 27–30

Rachel Donadio, 'A Museum Becomes a Battlefield Over Poland's History', *The New York Times* (New York, NY, 9 November 2016) <https://www.nytimes.com/2016/11/10/arts/design/museum-of-the-second-world-war-in-poland-debate.html> [accessed 23 June 2021]

Tomás Dorta, Edgar Pérez, and Annemarie Lesage, 'The Ideation Gap: Hybrid tools, Design Flow and Practice', *Design Studies*, 29.2 (2008), 121–141 <https://doi.org/10.1016/j.destud.2007.12.004>

Paul Dourish, *Where the Action Is: The Foundations of Embodied Interaction* (Cambridge, MA: The MIT Press, 2001)

Wilfred H. Drath, and Charles J. Palus, *Making Common Sense: Leadership as Meaning-Making in a Community of Practice* (Greensboro, NC: Center for Creative Leadership, 1994)

Carol Duncan, 'The Art Museum as Ritual', in *Civilizing Rituals* (London: Routledge, 1995) <https://doi.org/10.4324/9780203978719>

Abigail Durrant, Duncan Rowland, David S Kirk, Steve Benford, Joel E Fischer, and Derek McAuley, 'Automics: Souvenir Generating Photoware for Theme Parks', in *Proceedings of the SIGCHI Conference on Human Factors in Computing Systems* (ACM Press, 2011), pp. 1767–1776

EGMUS – European Group on Museum Statistics, 'EGMUS – Statistics' <https://www.egmus.eu/es/statistics/> [accessed 23 June 2021]

Lina Eklund, 'A Shoe Is a Shoe Is a Shoe: Interpersonalization and Meaning-making in Museums – Research Findings and Design Implications', *International Journal of Human-Computer Interaction*, 36.16 (2020), 1503–1513 <https://doi.org/10.1080/10447318.2020.1767982>

Brian Eno, *Ambient 1 – Music for Airports* (E.G. Records Ltd., 1978)

K Anders Ericsson, Ralf T Krampe, and Clemens Tesch-Römer, 'The Role of Deliberate Practice in the Acquisition of Expert Performance', *Psychological Review*, 100.3 (1993), 363–406

Europeana Digital Service Infrastructure Annual report (DSI-4) (Luxembourg: Europeana, 25 September 2020), p. 48 <https://pro.europeana.eu/post/europeana-dsi-4-annual-report-2019-2020> [accessed 4 June 2021]

Klint Finley, 'A Brief History of the End of the Comments', *Wired*, 8 October 2015 <https://www.wired.com/2015/10/brief-history-of-the-demise-of-the-comments-timeline/> [accessed 5 October 2020]

Martin Flintham, Stuart Reeves, Patrick Brundell, Tony Glover, Steve Benford, Duncan Rowland, and others, 'Flypad: Designing Trajectories in a Large-Scale Permanent Augmented Reality Installation', in *ECSCW 2011: Proceedings of the 12th European Conference on Computer Supported Cooperative Work, 24-28 September 2011, Aarhus Denmark*, ed. by Susanne Bødker, Niels Olof Bouvin, Volker Wulf, Luigina Ciolfi, and Wayne Lutters (London: Springer, 2011), pp. 233–252 <https://doi.org/10.1007/978-0-85729-913-0_13>

Lesley Fosh, Steve Benford, and Boriana Koleva, 'Supporting Group Coherence in a Museum Visit', in *Proceedings of the 19th ACM Conference on Computer-Supported Cooperative Work & Social Computing*, CSCW '16 (New York, NY, USA: ACM, 2016), pp. 1–12 <https://doi.org/10.1145/2818048.2819970>

Raghu Garud, Philipp Tuertscher, and Andrew H Van de Ven, 'Perspectives on Innovation Processes', *Academy of Management Annals*, 7.1 (2013), 775–819

William Gaver, John Bowers, Andrew Boucher, Hans Gellerson, Sarah Pennington, Albrecht Schmidt, and others, 'The Drift Table: Designing for Ludic

Engagement', in *CHI '04 Extended Abstracts on Human Factors in Computing Systems*, CHI EA '04 (New York, NY, USA: ACM, 2004), pp. 885–900 <https://doi.org/10.1145/985921.985947>

Mark S. Granovetter, 'The Strength of Weak Ties', *American Journal of Sociology*, 78.6 (1973), 1360–1380 <https://doi.org/10.1086/225469>

Eileen Hooper Greenhill, *Museums and the Shaping of Knowledge* (London: Routledge, 1992)

Kim Halskov, and Peter Dalsgaard, 'Inspiration Card Workshops', in *Proceedings of the 6th Conference on Designing Interactive Systems*, Proc. of Designing Interactive Systems (New York, NY, USA: ACM, 2006), pp. 2–11 <https://doi.org/10.1145/1142405.1142409>

Thomas S. Henricks, *Play and the Human Condition* (Champaign, IL: University of Illinois Press, 2015)

Thomas Duus Henriksen, 'Learning By Fiction', in *When Larp Grows Up: Theory and Methods in Larp*, ed. by Morten Gade, Line Thorup, and Mikkel Sander (Frederiksberg: Projektgruppen KP03, 2003), pp. 110–115 <https://nordiclarp.org/w/images/c/c2/2003-As.Larp.Grows.Up.pdf> [accessed 4 June 2021]

Hunter G. Hoffman, 'Physically Touching Virtual Objects Using Tactile Augmentation Enhances the Realism of Virtual Environments', in *Proceedings. IEEE 1998 Virtual Reality Annual International Symposium (Cat. No.98CB36180)* (presented at the Proceedings. IEEE 1998 Virtual Reality Annual International Symposium (Cat. No.98CB36180), Atlanta, GA: IEEE, 1998), pp. 59–63 <https://doi.org/10.1109/VRAIS.1998.658423>

Nanna Holdgaard, 'Online Museum Practices: A Holistic Analysis of Danish Museums and their Users' (unpublished PhD thesis, IT University of Copenhagen, 2014) <https://pure.itu.dk/portal/en/publications/online-museum-practices-a-holistic-analysis-of-danish-museums-and-their-users(13c275c8-616f-4203-91aa-e835049f5102).html> [accessed 4 June 2021]

Eilean Hooper-Greenhill, 'Audiences: A Curatorial Dilemma', in *Hooper-Greenhill, ed: The educational role of the museum.* (London & New York: Psychology Press, 1999), pp. 255–268

Eilean Hooper-Greenhill, 'Changing Values in the Art Museum: Rethinking Communication and Learning', *International Journal of Heritage Studies*, 6.1 (2000), 9–31 <https://doi.org/10.1080/135272500363715>

Eilean Hooper-Greenhill, *Museums and the Interpretation of Visual Culture*, Museum Meanings, 4 (Oxon, New York: Routledge, 2000)

Eilean Hooper-Greenhill, *Museums and the Shaping of Knowledge* (London & New York: Routledge, 1992)

Eva Hornecker, 'Creative Idea Exploration Within the Structure of a Guiding Framework: The Card Brainstorming Game', in *Proceedings of the Fourth International*

Conference on Tangible, Embedded, and Embodied Interaction, TEI '10 (New York, NY, USA: ACM, 2010), pp. 101–108 <https://doi.org/10.1145/1709886.1709905>

Eva Hornecker, and Jacob Buur, 'Getting a Grip on Tangible Interaction: A Framework on Physical Space and Social Interaction', in *Proceedings of the SIGCHI Conference on Human Factors in Computing Systems*, CHI '06 (New York, NY, USA: ACM, 2006), pp. 437–446 <https://doi.org/10.1145/1124772.1124838>

Eva Hornecker, and Luigina Ciolfi, 'Human-Computer Interactions in Museums', *Synthesis Lectures on Human-Centered Informatics*, 12.2 (2019), i–171 <https://doi.org/10.2200/S00901ED1V01Y201902HCI042>

S. Hsi, 'A Study of User Experiences Mediated by Nomadic Web Content in a Museum', *Journal of Computer Assisted Learning*, 19.3 (2003), 308–319 <https://doi.org/10.1046/j.0266-4909.2003.jca_023.x>

Johan Huizinga, *Homo Ludens: A Study of the Play-elememt in Culture* (Angelico Press (Reprint 2016), 1955)

Brent Edward Insko, 'Passive Haptics Significantly Enhances Virtual Environments' (unpublished PhD Thesis, The University of North Carolina at Chapel Hill, 2001) <https://dl.acm.org/doi/10.5555/933178>

P. Ioannidis, L. Eklund, and Anders Sundnes Løvlie, 'A Lifecycle Study of a Substitutional Reality Installation in a Museum Space', *Journal on Computing and Cultural Heritage.*, Forthcoming (2020)

Katherine Isbister, *How Games Move Us: Emotion by Design*, Playful Thinking (MIT Press, 2017)

Jan Inge Jenssen, and Geir Jørgensen, 'How Do Corporate Champions Promote Innovations?', *International Journal of Innovation Management*, 8.01 (2004), 63–86 <https://doi.org/10.1142/S1363919604000964>

Ben Jonson, 'Design Ideation: The Conceptual Sketch in the Digital Age', *Design Studies*, 26.6 (2005), 613–624

David Kadish, Sebastian Risi, and Anders Sundnes Løvlie, 'Improving Object Detection in Art Images Using Only Style Transfer', in *The International Joint Conference on Neural Networks (IJCNN)* (presented at the The International Joint Conference on Neural Networks (IJCNN), Virtual Event: IEEE, 2021) <https://arxiv.org/abs/2102.06529> [accessed 23 June 2021]

Line Vestergaard Knudsen, and Anne Rørbæk Olesen, 'Complexities of collaborating: Understanding and Managing Differences in Collaborative Design of Museum Communication', in *The Routledge Handbook of Museums, Media and Communication*, ed. by Kirsten Drotner, Vince Dziekan, Ross Parry, and Kim Christian Schrøder (London and New York: Routledge, 2019), pp. 205–218 <https://gifting.digital/wp-content/uploads/2020/01/Preprint_complexities-of-collaborating-1.pdf> [accessed 29 October 2019]

Tarja-Kaarina Laamanen, and Pirita Seitamaa-Hakkarainen, 'Interview Study of Professional Designers' Ideation Approaches', *The Design Journal*, 17.2 (2014), 194–217 <https://doi.org/10.2752/175630614X13915240575988>

Jean Lave, and Etienne Wenger, *Situated Learning: Legitimate Peripheral Participation* (Cambridge U.K.: Cambridge university press, 1991)

Kurt Lewin, 'Action Research and Minority Problems', *Journal of Social Issues*, 2.4 (1946), 34–46 <https://doi.org/10.1111/j.1540-4560.1946.tb02295.x>

Gunnar Liestøl, 'PowerPoint: Beyond Hardware and Software', in *Inside Multimodal Composition*, ed. by Andrew Morrison (New York: Hampton Press, 2009), pp. 71–93

Anders Sundnes Løvlie, Edgar Bodiaj, Karin Ryding, Steve Benford, Harriet Cameron, Dimitrios Darzentas, and others, *Report on Sensitive Pictures*, The GIFT Project (Brussels, Belgium: The Community Research and Development Information Service (CORDIS), The European Commission, 20 January 2020), p. 52 <https://ec.europa.eu/research/participants/documents/downloadPublic?documentIds=080166e5cb4633ce&appId=PPGMS> [accessed 4 June 2021]

Anders Sundnes Løvlie, Lina Eklund, Annika Waern, Karin Ryding, and Paulina Rajkowska, 'Designing for Interpersonal Museum Experiences', in *Museums and the Challenge of Change: Old Institutions in a New World*, ed. by Graham Black (London & New York: Routledge, 2020), pp. 223–238

Jonas Löwgren, and Erik Stolterman, *Thoughtful Interaction Design: A Design Perspective on Information Technology* (Cambridge, MA: Mit Press, 2004)

Ioanna Lykourentzou, Xavier Claude, Yannick Naudet, Eric Tobias, Angeliki Antoniou, George Lepouras, and others, 'Improving Museum Visitors' Quality of Experience through Intelligent Recommendations: A Visiting Style-Based Approach', *Workshop Proceedings of the 9th International Conference on Intelligent Environments*, 2013, 507–518 <https://doi.org/10.3233/978-1-61499-286-8-507>

Bernadette T Lynch, 'Custom-Made Reflective Practice: Can Museums Realise their Capabilities in Helping Others Realise Theirs?', *Museum Management and Curatorship*, 26.5 (2011), 441–458

Leilah Lyons, 'Designing Opportunistic User Interfaces to Support a Collaborative Museum Exhibit', in *Proceedings of the 9th International Conference on Computer Supported Collaborative Learning*, CSCL'09 (Rhodes, Greece: International Society of the Learning Sciences, 2009), I, 375–384

Sharon Macdonald, *A Companion to Museum Studies* (John Wiley & Sons, 2011), XXXIX

Sharon Macdonald, and Paul Basu, *Exhibition Experiments* (John Wiley & Sons, 2008)

Stephann Makri, and Ann Blandford, 'What is Serendipity? A Workshop Report', *Information Research*, 16.3 (2011) <http://www.informationr.net/ir/16-3/paper491.html> [accessed 25 June 2020]

Didier Maleuvre, *Museum Memories: History, Technology, Art* (Stanford University Press, 1999)

Bronislaw Malinowski, *Argonauts of the Western Pacific: An Account of Native Enterprise and Adventure in the Archipelagoes of Melanesian New Guinea* (London: Routledge (reprint 1978), 1922)

Elena Márquez Segura, Annika Waern, Luis Márquez Segura, and David López Recio, 'Playification: The PhySeEar Case', in *Proceedings of the 2016 Annual Symposium on Computer-Human Interaction in Play* (presented at the Proceedings of the 2016 annual symposium on computer-human interaction in play, ACM, 2016), pp. 376–388 <https://doi.org/10.1145/2967934.2968099>

Nikita Mathias, 'From Website Discussions to Transforming the Organisation', *The GIFT Box.*, 2019 <https://gifting.digital/munch-museum/> [accessed 4 June 2021]

Friedemann Mattern, and Christian Floerkemeier, 'From the Internet of Computers to the Internet of Things', in *From Active Data Management to Event-Based Systems and More: Papers in Honor of Alejandro Buchmann on the Occasion of His 60th Birthday*, ed. by Kai Sachs, Ilia Petrov, and Pablo Guerrero, Lecture Notes in Computer Science (Berlin and Heidelberg: Springer, 2010), pp. 242–259 <https://doi.org/10.1007/978-3-642-17226-7_15>

Marcel Mauss, *The Gift: Forms and Functions of Exchange in Archaic Societies* (London: W. W. Norton & Company (reprint 2000), 1925)

John C Maxwell, *Failing Forward: How to Make the Most of Your Mistakes* (Thomas Nelson Publishers, 2000)

Jane Evelyn McGonigal, *Reality Is Broken: Why Games Make Us Better and How They Can Change the World* (Penguin Press HC, The, 2010) – incomplete publication details

P. McManus, 'What People Say and How They Think in a Science Museum', in *Heritage Interpretation, Vol 2: The Visitor Experience*, ed. by D.L. Uzzell (London and New York: Bellhaven Press, 1989), pp. 156–165

Nick Merriman, 'The Crisis of Representation in Archaeological Museums', in *Cultural Resource Management in Contemporary Society*, ed. by *Hatton and MacManamon* (Routledge, 2000), pp. 316–325

Victor T.C. Middleton, 'The Future Demand for Museums 1990-2001', in *The Museums Profession: Internal and External Relations*, ed. by Gaynor Kavanagh, Leicester Museum Studies Series (Leicester, UK: Leicester University Press, 1991), pp. 139–160 <https://www.abebooks.co.uk/Museums-Profession-Internal-External-Relations-Leicester/22671512755/bd> [accessed 29 October 2019]

Paul Milgram, and Fumio Kishino, 'A Taxonomy of Mixed Reality Visual Displays', *IEICE TRANSACTIONS on Information and Systems*, E77-D.12 (1994), 1321–1329 <https://search.ieice.org/bin/summary.php?id=e77-d_12_1321> [accessed 25 February 2019]

Jay Mollica, 'Send Me SFMOMA', *San Francisco Museum of Modern Art*, 2017 <https://www.sfmoma.org/read/send-me-sfmoma/> [accessed 23 September 2019]

Mark Molloy, 'The Real Story Behind a Viral Rembrandt "Kids on Phones" Photo', 16 January 2016, section News <https://www.telegraph.co.uk/news/newstopics/howaboutthat/12103150/Rembrandt-The-Night-Watch-The-real-story-behind-the-kids-on-phones-photo.html> [accessed 21 February 2020]

Markus Montola, Jaakko Stenros, and Annika Waern, *Pervasive Games: Theory and Design* (Boca Raton, FL: CRC Press, 2009)

Eduardo Mortimer, and Philip Scott, *Meaning Making in Secondary Science Classrooms*, 1st edn (USA: McGraw-Hill, Inc., 2003)

Scott Nicholson, 'A RECIPE for Meaningful Gamification', in *Gamification in Education and Business*, ed. by Torsten Reiners and Lincoln C. Wood (Cham: Springer International Publishing, 2015), pp. 1–20 <https://doi.org/10.1007/978-3-319-10208-5_1>

Niki Nikonanou, and Foteini Venieri, 'Museums as Gameworlds: The Use of Live Action Role Playing Games in Greek Museums', *International Journal of the Inclusive Museum*, 6.3 (2014)

Elena Not, and Daniela Petrelli, 'Blending Customisation, Context-Awareness and Adaptivity for Personalised Tangible Interaction in Cultural Heritage', *International Journal of Human-Computer Studies*, Advanced User Interfaces for Cultural Heritage, 114 (2018), 3–19 <https://doi.org/10.1016/j.ijhcs.2018.01.001>

Anne Rørbæk Olesen, 'For the Sake of Technology? The Role of Technology Views in Funding and Designing Digital Museum Communication', *Museum Management and Curatorship*, 31.3 (2016), 283–298

Anne Rørbæk Olesen, 'Mapping Innovation Processes: Visual Techniques for Opening and Presenting the Black Box of Service Innovation Processes', in *Research Methods in Service Innovation* (Edward Elgar Publishing, 2017) <http://www.e-elgar.com/shop/research-methods-in-service-innovation> [accessed 4 June 2021]

Anne Rørbæk Olesen, Nanna Holdgaard, and Ditte Laursen, 'Challenges of Practicing Digital Imaginaires in Collaborative Museum Design', *CoDesign*, 0.0 (2018), 1–13 <https://doi.org/10.1080/15710882.2018.1539109>

Anne Rørbæk Olesen, Nanna Holdgaard, and Anders Sundnes Løvlie, 'Co-Designing a Co-Design Tool to Strengthen Ideation in Digital Experience Design at Museums', *CoDesign*, 0.0 (2020), 1–16 <https://doi.org/10.1080/15710882.2020.1812668>

Antti Oulasvirta, Esko Kurvinen, and Tomi Kankainen, 'Understanding Contexts by Being There: Case Studies in Bodystorming', *Personal and Ubiquitous Computing*, 7.2 (2003), 125–134 <https://doi.org/10.1007/s00779-003-0238-7>

Ross Parry, *Recoding the Museum: Digital Heritage and the Technologies of Change* (New York: Routledge, 2007)

Darren Peacock, 'Making Ways for Change: Museums, Disruptive Technologies and Organisational Change', *Museum Management and Curatorship*, 23.4 (2008), 333–351
Susan M Pearce, *Museums and the Appropriation of Culture* (The Athlone Press, 1994)
Andrew Pekarik, Zahava D. Doering, and David A. Karns, 'Exploring Satisfying Experiences in Museums', *Curator: The Museum Journal*, 42.2 (1999), 152–173
Jon Pemberton, Sharon Mavin, and Brenda Stalker, 'Scratching Beneath the Surface of Communities of (Mal)Practice', *The Learning Organization*, 2007
Daniela Petrelli, Luigina Ciolfi, Dick van Dijk, Eva Hornecker, Elena Not, and Albrecht Schmidt, 'Integrating Material and Digital: A New Way for Cultural Heritage', *Interactions*, 20.4 (2013), 58–63 <https://doi.org/10.1145/2486227.2486239>
Daniela Petrelli, Mark T. Marshall, Sinéad O'Brien, Patrick McEntaggart, and Ian Gwilt, 'Tangible Data Souvenirs as a Bridge Between a Physical Museum Visit and Online Digital Experience', *Personal and Ubiquitous Computing*, 21.2 (2017), 281–295 <https://doi.org/10.1007/s00779-016-0993-x>
Emily Pringle, 'Developing the Practitioner-Researcher Within the Art Museum', *Tate Papers*, 29, 2018, 16–17
Stefan Rennick-Egglestone, Patrick Brundell, Boriana Koleva, Steve Benford, Maria Roussou, and C. Chaffardon, 'Families and Mobile Devices in Museums: Designing for Integrated Experiences', *Journal on Computing and Cultural Heritage*, 9.2 (2016), 11:1-11:13 <https://doi.org/10.1145/2891416>
Ed Rodley, 'Playing With the Past, Part Two: Magic Circles and Interaction Alibis', *Thinking about Museums*, 2018 <https://thinkingaboutmuseums.com/2018/03/12/playing-with-the-past-part-two-magic-circles-and-interaction-alibis/> [accessed 16 June 2021]
Everett Rogers, *Diffussion of Innovations*, 5th edition (New York: Simon & Schuster, 2003)
Harold Rosen, 'The Importance of Story', *Language Arts*, 63.3 (1986), 226–237 <https://www.jstor.org/stable/41405427> [accessed 16 June 2021]
Ben Russell, 'Headmap Manifesto' (headmap.org, 1999) <http://technoccult.net/wp-content/uploads/library/headmap-manifesto.pdf> [accessed 2 November 2009]
Julie A. Ruth, Cele C. Otnes, and Frederic F. Brunel, 'Gift Receipt and the Reformulation of Interpersonal Relationships', *Journal of Consumer Research*, 25.4 (1999), 385–402 <https://doi.org/10.1086/209546>
Karin Ryding, 'The Silent Conversation: Designing for Introspection and Social Play in Art Museums', in *Proceedings of the 2020 CHI Conference on Human Factors in Computing Systems*, CHI '20 (Honolulu, HI, USA: Association for Computing Machinery, 2020), pp. 1–10 <https://doi.org/10.1145/3313831.3376357>
Karin Ryding, Jocelyn Spence, Anders Sundnes Løvlie, and Steve Benford, 'Interpersonalizing Intimate Museum Experiences', *International Journal of*

Human–Computer Interaction, 2021, 1–22 <https://doi.org/10.1080/10447318.2020.1870829>

Richard Sagor, *How to Conduct Collaborative Action Research* (Alexandria, VA: Association for Supervision and Curriculum Development, 1993)

Elizabeth B.-N. Sanders, and Pieter Jan Stappers, 'Probes, Toolkits and Prototypes: Three Approaches to Making in Codesigning', *CoDesign*, 10.1 (2014), 5–14 <https://doi.org/10.1080/15710882.2014.888183>

Dennis Schleicher, Peter Jones, and Oksana Kachur, 'Bodystorming as Embodied Designing', *Interactions*, 17.6 (2010), 47–51 <https://doi.org/10.1145/1865245.1865256>

Donald A. Schön, *Beyond the Stable State* (New York: Random House, 1971)

Donald A. Schön, *The Reflective Practitioner: How Professionals Think in Action* (Aldershot, UK: Ashgate, 1983)

Barry Schwartz, 'The Social Psychology of the Gift', *American Journal of Sociology*, 73.1 (1967), 1–11 <https://www.jstor.org/stable/2776124> [accessed 16 June 2021]

Werner Schweibenz, 'The Virtual Museum: An Overview of its Origins, Concepts, and Terminology', *The Museum Review*, 4.1 (2019), 2–28 <https://themuseumreviewjournal.wordpress.com/2019/08/02/tmr_vol4no1_schweibenz/>

Peter M Senge, *The Fifth Discipline: The Art and Practice of the Learning Organization*, 2nd edn (New York: Doubleday/Currency, 2006)

Natalya Sergeeva, 'What Makes an "Innovation Champion"?', *European Journal of Innovation Management*, 19.1 (2016), 72–89 <https://doi.org/10.1108/EJIM-06-2014-0065>

Beverly Serrell, 'Paying Attention: The Duration and Allocation of Visitors' Time in Museum Exhibitions', *Curator: The Museum Journal*, 40.2 (1997), 108–125 <https://doi.org/10.1111/j.2151-6952.1997.tb01292.x>

Mahima Sharma, Sudhanshu Kacker, and Mohit Sharma, 'A Brief Introduction and Review on Galvanic Skin Response', 2.6 (2016), 13–17 <https://doi.org/10.21276/ijmrp.2016.2.6.003>

Daniel J. Sherman, 'Quatremère/Benjamin/Marx: Art Museums, Aura, and Commodity Fetishism', in *Museum Culture: Histories, Discourses, Spectacles*, ed. by Itit Rogoff and Daniel J. Sherman (London: Routledge, 1994), pp. 143–163 <https://doi.org/10.4324/9780203168196>

Kathy G. Short, 'Story as World Making', *Language Arts*, 90.1 (2012), 9–17

Miguel Sicart, *Beyond Choices: The Design of Ethical Gameplay* (Cambridge, MA: The MIT Press, 2013)

Miguel Sicart, 'Toward an Ethics of Homo Ludens', in *Ludics: Play as Humanistic Inquiry*, ed. by Vassiliki Rapti and Eric Gordon (Singapore: Springer, 2021), pp. 21–45 <https://doi.org/10.1007/978-981-15-7435-1_2>

Lois H. Silverman, *The Social Work of Museums* (London & New York: Routledge, 2010)

Lois H. Silverman, 'Visitor Meaning-Making in Museums for a New Age', *Curator: The Museum Journal*, 38.3 (1995), 161–170 <https://doi.org/10.1111/j.2151-6952.1995.tb01052.x>

Adalberto L. Simeone, Eduardo Velloso, and Hans Gellersen, 'Substitutional Reality: Using the Physical Environment to Design Virtual Reality Experiences', in *Proceedings of the 33rd Annual ACM Conference on Human Factors in Computing Systems*, CHI '15 (New York, NY, USA: Association for Computing Machinery, 2015), pp. 3307–3316 <https://doi.org/10.1145/2702123.2702389>

Nina Simon, *The Participatory Museum* (Santa Cruz, California: Museum 2.0, 2010) <http://www.participatorymuseum.org/> [accessed 27 January 2016]

Lisa F. Smith, Jeffrey K. Smith, and Pablo P. L. Tinio, 'Time Spent Viewing Art and Reading Labels', *Psychology of Aesthetics Creativity and the Arts*, 11.1 (2017), 77–85 <https://psycnet.apa.org/doi/10.1037/aca0000049>

Kingkarn Sookhanaphibarn, and Ruck Thawonmas, 'A Movement Data Analysis and Synthesis Tool for Museum Visitors' Behaviors', in *Advances in Multimedia Information Processing – PCM 2009*, ed. by Paisarn Muneesawang, Feng Wu, Itsuo Kumazawa, Athikom Roeksabutr, Mark Liao, and Xiaoou Tang, Lecture Notes in Computer Science (Berlin, Heidelberg: Springer, 2009), pp. 144–154 <https://doi.org/10.1007/978-3-642-10467-1_12>

Barbara J. Soren, 'Museum Experiences that Change Visitors', *Museum Management and Curatorship*, 24.3 (2009), 233–251 <https://doi.org/10.1080/09647770903073060>

Adriana de Souza e Silva, 'From Cyber to Hybrid: Mobile Technologies as Interfaces of Hybrid Spaces', *Space and Culture*, 9.3 (2006), 261–278 <https://doi.org/10.1177/1206331206289022>

Jocelyn Spence, 'Inalienability: Understanding Digital Gifts', in *Proceedings of the 2019 CHI Conference on Human Factors in Computing Systems*, CHI '19 (New York, NY, USA: ACM, 2019), p. 657:1-657:12 <https://doi.org/10.1145/3290605.3300887>

Jocelyn Spence, 'Using Hybrid Gifting to Build Personal Engagement in and with Museums' (presented at the Museums Matter: Interpreting, Sharing and Collaborating in Research, Nottingham, UK, 2019) <https://www.nottingham.ac.uk/home/featureevents/2019/museums-matter-%E2%80%94-interpreting-sharing-and-collaborating-in-research%E2%80%94.aspx>

Jocelyn Spence, Benjamin Bedwell, Michelle Coleman, Steve Benford, Boriana N. Koleva, Matt Adams, and others, 'Seeing with New Eyes: Designing for In-the-Wild Museum Gifting', in *Proceedings of ACM CHI Conference on Human Factors in Computing Systems (CHI 2019)* (presented at the CHI 2019, Glasgow, Scotland, UK: ACM, 2019) <https://doi.org/10.1145/3290605.3300235>

Jocelyn Spence, Dimitrios Paris Darzentas, Yitong Huang, Harriet R. Cameron, Eleanor Beestin, and Steve Benford, 'VRtefacts: Performative Substitutional

Reality with Museum Objects', in *Proceedings of the 2020 ACM Designing Interactive Systems Conference*, DIS '20 (New York, NY, USA: Association for Computing Machinery, 2020), pp. 627–640 <https://doi.org/10.1145/3357236.3395459>

Susan Leigh Star, 'This is Not a Boundary Object: Reflections on the Origin of a Concept', *Science, Technology, & Human Values*, 35.5 (2010), 601–617 <https://doi.org/10.1177/0162243910377624>

Susan Leigh Star, and James R. Griesemer, 'Institutional Ecology, `Translations' and Boundary Objects: Amateurs and Professionals in Berkeley's Museum of Vertebrate Zoology, 1907-39', *Social Studies of Science*, 19.3 (1989), 387–420 <https://doi.org/10.1177/030631289019003001>

Jakko Stenros, 'In Defence of a Magic Circle: The Social, Mental and Cultural Boundaries of Play', in *2012 International DiGRA Nordic Conference* (presented at the DiGRA Nordic '12, Tampere, Finland: DiGRA, 2012), x, 1–19

Linda Stoltze, Tim Wray, and Anders Sundnes Løvlie, 'Reflect in One Minute', *The GIFT Box* <https://gifting.digital/one-minute-experience/> [accessed 23 June 2021]

Dagny Stuedahl, 'Participation in Design and Changing Practices of Museum Development', in *The Routledge Handbook of Museums, Media and Communication*, ed. by Kirsten Drotner, Vince Dziekan, Ross Parry, and Kim Christian Schrøder, Routledge International Handbooks (London and New York: Routledge, 2019), pp. 218–231

Gerald I. Susman, and Roger D. Evered, 'An Assessment of the Scientific Merits of Action Research', *Administrative Science Quarterly*, 23.4 (1978), 582–603 <https://doi.org/10.2307/2392581>

Trine Syvertsen, and Gunn Enli, 'Digital Detox: Media Resistance and the Promise of Authenticity', *Convergence*, 26.5–6 (2020), 1269–1283 <https://doi.org/10.1177/1354856519847325>

Steven Taylor, 'Leaving the Field: Research, Relationships, and Responsibilities', in *Experiencing Fieldwork: An Inside View of Qualitative Research*, 124 vols. (Thousand Oaks: SAGE Publications, Inc., 1991), pp. 238–247 <https://doi.org/10.4135/9781483325514>

Paul Tennent, Sarah Martindale, Steve Benford, Dimitrios Darzentas, Pat Brundell, and Mat Collishaw, 'Thresholds: Embedding Virtual Reality in the Museum', *Journal on Computing and Cultural Heritage*, 13.2 (2020), 12:1-12:35 <https://doi.org/10.1145/3369394>

Gillian Tett, *The Silo Effect: The Peril of Expertise and the Promise of Breaking Down Barriers* (New York: Simon and Schuster, 2015)

Peter Tolmie, Steve Benford, Chris Greenhalgh, Tom Rodden, and Stuart Reeves, 'Supporting Group Interactions in Museum Visiting', in *Proceedings of the 17th ACM Conference on Computer Supported Cooperative Work & Social Computing*, CSCW '14 (New York, NY, USA: ACM, 2014), pp. 1049–1059 <https://doi.org/10.1145/2531602.2531619>

Colwyn Trevarthen, 'The Generation of Human Meaning: How Shared Experience Grows in Infancy', in *Joint Attention: New Developments in Psychology, Philosophy of Mind, and Social Neuroscience*, ed. by A. Seemann (Cambridge, MA, US: MIT Press, 2011), pp. 73–113

Haridimos Tsoukas, and Robert Chia, 'On Organizational Becoming: Rethinking Organizational Change', *Organization Science*, 13.5 (2002), 567–582 <https://doi.org/10.1287/orsc.13.5.567.7810>

Sherry Turkle, *Alone Together: Why We Expect More from Technology and Less from Each Other* (New York: Basic Books, 2011) <https://www.basicbooks.com/titles/sherry-turkle/alone-together/9780465093656/> [accessed 23 June 2021]

Marc Tuters, and Kazys Varnelis, 'Beyond Locative Media: Giving Shape to the Internet of Things', *Leonardo*, 39.4 (2006), 357–363 <https://doi.org/10.1162/leon.2006.39.4.357>

Vasiliki Tzibazi, 'Participatory Action Research with Young People in Museums', *Museum Management and Curatorship*, 28.2 (2013), 153–171 <https://doi.org/10.1080/09647775.2013.776800>

Bjarki Valtysson, 'Europeana: The Digital Construction of Europe's Collective Memory', *Information, Communication & Society*, 15.2 (2012), 151–170 <https://doi.org/10.1080/1369118X.2011.586433>

Andrew H. Van de Ven, Douglas Polley, Raghu Garud, and Sankaran Venkataraman, *The Innovation Journey* (Oxford, UK: Oxford University Press, 2008)

Annika Waern, Paulina Rajkowska, Karin B. Johansson, Jon Bac, Jocelyn Spence, and Anders Sundnes Løvlie, 'Sensitizing Scenarios: Sensitizing Designer Teams to Theory', in *Proceedings of the 2020 CHI Conference on Human Factors in Computing Systems*, CHI '20 (New York, NY, USA: Association for Computing Machinery, 2020), pp. 1–13 <https://doi.org/10.1145/3313831.3376620>

Max Weber, 'The Nature of Social Action', in *Max Weber: Selections in Translation*, ed. by W. G. Runciman, trans. by E. Matthews (Cambridge: Cambridge University Press, 1978), pp. 7–32 <https://doi.org/10.1017/CBO9780511810831.005>

Annette B. Weiner, *Inalienable Possessions: The Paradox of Keeping-While Giving* (Oakland, CA: University of California Press, 1992) <https://www.jstor.org/stable/10.1525/j.ctt1ppcbw> [accessed 23 June 2021]

Mark Weiser, 'The Computer for the 21st Century', *SIGMOBILE Mob. Comput. Commun. Rev.*, 3.3 (1999), 3–11 <https://doi.org/10.1145/329124.329126>

Carol H. Weiss, 'Nothing as Practical as Good Theory: Exploring Theory-Based Evaluation for Comprehensive Community Initiatives for Children and Families', in *New Approaches to Evaluating Community Initiatives: Concepts, Methods, and Contexts*, ed. by James P. Connell, Anne C. Kubisch, Lisbeth B. Schorr, and Carol H. Weiss (Washington DC: Aspen Institute, 1995), pp. 65–92

Etienne Wenger, *Communities of Practice: Learning, Meaning, and Identity*, Learning in Doing: Social, Cognitive and Computational Perspectives (Cambridge: Cambridge University Press, 1998) <https://doi.org/10.1017/CBO9780511803932>

Richard Wetzel, Tom Rodden, and Steve Benford, 'Developing Ideation Cards for Mixed Reality Game Design', *Transactions of the Digital Games Research Association*, 3.2 (2017) <https://doi.org/10.26503/todigra.v3i2.73>

Andrea Witcomb, 'The Materiality of Virtual Technologies: A New Approach to Thinking about the Impact of Multimedia in Museums', in *Theorizing Digital Cultural Heritage: A Critical Discourse*, ed. by Fiona Cameron and Sarah Kenderdine (Cambridge, MA: The MIT Press, 2007), pp. 35–48 <https://mitpress.universitypressscholarship.com/view/10.7551/mitpress/9780262033534.001.0001/upso-9780262033534-chapter-3> [accessed 23 June 2021]

Allison Woodruff, Paul Aoki, Amy Hurst, and Margaret Szymanski, 'Electronic Guidebooks and Visitor Attention', in *International Cultural Heritage Informatics Meeting: Proceedings from ichim01*, ed. by D. Bearman and F. Garzotto (presented at the International Cultural Heritage Informatics Meeting, Milano, Italy: Archives & Museum Informatics, 2001) <http://www.archimuse.com/publishing/ichim01_vol1/woodruff.pdf>

Sarah Younas, 'From Failure Cafés to Rapid Prototyping', *The GIFT Box*, 2019 <https://gifting.digital/tyne-and-wear/> [accessed 2 April 2020]

Shaker Zahra, James Hayton, Jeremy Marcel, and Hugh O'Neill, 'Fostering Entrepreneurship During International Expansion: Managing Key Challenges', *European Management Journal*, 19.4 (2001), 359–369 <https://doi.org/10.1016/S0263-2373(01)00037-8>

Monika Żychlińska, and Erica Fontana, 'Museal Games and Emotional Truths: Creating Polish National Identity at the Warsaw Rising Museum', *East European Politics and Societies*, 30.2 (2016), 235–269 <https://doi.org/10.1177/0888325414566198>

Index

For Product Safety Concerns and Information please contact our EU representative GPSR@taylorandfrancis.com Taylor & Francis Verlag GmbH, Kaufingerstraße 24, 80331 München, Germany